I0083854

Gender and Early Television

Library of Gender and Popular Culture

From *Mad Men* to gaming culture, performance art to steampunk fashion, the presentation and representation of gender continues to saturate popular media. This series seeks to explore the intersection of gender and popular culture, engaging with a variety of texts – drawn primarily from Art, Fashion, TV, Cinema, Cultural Studies and Media Studies – as a way of considering various models for understanding the complementary relationship between 'gender identities' and 'popular culture'. By considering race, ethnicity, class and sexual identities across a range of cultural forms, each book in the series adopts a critical stance towards issues surrounding the development of gender identities and popular and mass cultural 'products'.

For further information or enquiries,
please contact the library series editors:

Claire Nally: claire.nally@northumbria.ac.uk
Angela Smith: angela.smith@sunderland.ac.uk

Advisory Board:
Dr Kate Ames, Central Queensland University, Australia
Dr Michael Higgins, University of Strathclyde, UK
Prof Åsa Kroon, Örebro University, Sweden
Dr Andrea McDonnell, Emmanuel College, USA
Dr Niall Richardson, University of Sussex, UK
Dr Jacki Willson, University of Leeds, UK

Library of Gender
& Popular Culture

Published and forthcoming titles:

The Aesthetics of Camp: Post-Queer Gender and Popular Culture
By Anna Malinowska

Ageing Femininity on Screen: The Older Woman in Contemporary Cinema
By Niall Richardson

All-American TV Crime Drama: Feminism and Identity Politics in Law and Order: Special Victims Unit
By Sujata Moorti and Lisa Cuklanz

Bad Girls, Dirty Bodies: Sex, Performance and Safe Femininity
By Gemma Commane

Beyoncé: Celebrity Feminism in the Age of Social Media
By Kirsty Fairclough-Isaacs

Conflicting Masculinities: Men in Television Period Drama
By Katherine Byrne, Julie Anne Taddeo and James Leggott (Eds)

Fat on Film: Gender, Race and Body Size in Contemporary Hollywood Cinema
By Barbara Plotz

Fathers on Film: Paternity and Masculinity in 1990s Hollywood
By Katie Barnett

Film Bodies: Queer Feminist Encounters with Gender and Sexuality in Cinema
By Katharina Lindner

Gay Pornography: Representations of Sexuality and Masculinity
By John Mercer

Gender and Austerity in Popular Culture: Femininity, Masculinity and Recession in Film and Television
By Helen Davies and Claire O'Callaghan (Eds)

The Gendered Motorcycle: Representations in Society, Media and Popular Culture
By Esperanza Miyake

Gendering History on Screen: Women Filmmakers and Historical Films
By Julia Erhart

Girls Like This, Boys Like That: The Reproduction of Gender in Contemporary Youth Cultures
By Victoria Cann

The Gypsy Woman: Representations in Literature and Visual Culture
By Jodie Matthews

Gender and Early Television

Mapping Women's Role in Emerging US and British Media, 1850–1950

Sarah Arnold

BLOOMSBURY ACADEMIC
LONDON • NEW YORK • OXFORD • NEW DELHI • SYDNEY

BLOOMSBURY ACADEMIC
Bloomsbury Publishing Plc
50 Bedford Square, London, WC1B 3DP, UK
1385 Broadway, New York, NY 10018, USA
29 Earlsfort Terrace, Dublin 2, Ireland

BLOOMSBURY, BLOOMSBURY ACADEMIC and the Diana logo
are trademarks of Bloomsbury Publishing Plc

First published in Great Britain 2021
Paperback edition published 2023

Copyright © Sarah Arnold, 2021, 2023

Sarah Arnold has asserted her right under the Copyright, Designs
and Patents Act, 1988, to be identified as Author of this work.

For legal purposes the Acknowledgements on p. xii constitute
an extension of this copyright page.

Cover design: Charlotte Daniels
Television director Frances Buss, hostess of *Vanity Fair*,
Dorothy Doan, and her assistant Anne Kelleher talking on the set of *Vanity Fair*,
1st October 1948 (© CBS Photo Archive / Getty Images)

All rights reserved. No part of this publication may be reproduced or transmitted
in any form or by any means, electronic or mechanical, including photocopying,
recording, or any information storage or retrieval system, without prior
permission in writing from the publishers.

Bloomsbury Publishing Plc does not have any control over, or responsibility for,
any third-party websites referred to or in this book. All internet addresses given
in this book were correct at the time of going to press. The author and publisher
regret any inconvenience caused if addresses have changed or sites have ceased
to exist, but can accept no responsibility for any such changes.

A catalogue record for this book is available from the British Library.

A catalog record for this book is available from the Library of Congress.

ISBN: HB: 978-1-7807-6976-9
        PB: 978-1-3502-4007-0
        ePDF: 978-1-7867-3616-1
        eBook: 978-1-7867-2610-0

Series: Library of Gender and Popular Culture

Typeset by Integra Software Services Pvt. Ltd.

To find out more about our authors and books visit www.bloomsbury.com
and sign up for our newsletters

Contents

Illustrations

Series Editors' Foreword

The Library of Gender and Popular Culture offers insights into gender and sexuality in a wide range of interpretations of 'popular culture'. Whilst much of this relates to the representation of gender, the importance of the production of such media and cultures is less well explored. A notable exception in this series is Julia Erhart's *Gendering History on Screen* (2018), and so Sarah Arnold's book is a welcome development of the arguments that emerged in Erhart's work in relation to women as producers. Although television might appear to have a shorter history than film, as Arnold explores here, this history is actually much longer than the actual arrival of television for the masses in the 1950s. Developments in technology that lead to television start in the late nineteenth century at a time when women's rights were just beginning to be raised in both the UK and United States. The 'New Woman' emerged and engaged with innovations, but, by virtue of her lack of technical education, was largely excluded from the sphere of production. As Arnold shows, the great world wars of the early part of the twentieth century curtailed the development of television, and it was not until the 1950s that the industry resumed, a time that coincides with the retraditionalization of gender roles in the immediate aftermath of the Second World War. That said, one of the hidden histories that Arnold reveals is that women were involved in experimental television in the 1930s, as producers and presenters. By contrast, post-war, women were seen as consumers of television. However, women are not simply excluded because they don't have access to the skills and technical knowledge, but because the culture and language of that technology are inherently masculine. Women were therefore largely excluded from television production both technically and culturally. Instead, their perceived place as consumers of television, as Arnold explains, was a role that had largely been defined for them by television audience research, offering a limited set of roles: wife, mother and carer.

She also points to how television audience research produced rather than reflected the female audience. However, this producer/audience dichotomy was not always so. Arnold's analysis takes us into the hidden corners of television production, where women can be found throughout its history. These hidden women in the shadows of popular culture are typical of the women found in many of the books in this series, and as such this book offers an engaging and pertinent link with the theme of gender in popular culture, by revealing the lesser-known aspects of women in the production of popular culture.

Acknowledgements

I am very grateful for the support of the various archives, museums and funding bodies that enabled my research for this book. I would like to extend my thanks to the following for helping me to source information that was crucial to the development of this project: Steve Jajkowski, Archives Director at the Museum of Broadcast Communication in Chicago; Laura Schnitker and Michael Henry with the Mass Media & Culture collections at the University of Maryland; Katie Ankers and the staff at the BBC Written Archives Centre in Caversham; Jane Klain of The Paley Center for Media; and Lee Grady of the Wisconsin Historical Society. I am also very appreciative of the research and financial support offered by the following: Bríd Dooley and members of the International Federation of Television Archives for awarding me the Media Studies Grant; and Joshua Larkin Rowley and the staff at the Rubenstein Library at Duke University for their help and for offering me the John Furr Fellowship and Travel Grant at Duke University.

I would also like to extend my thanks to colleagues at Maynooth University who have supported this project during its years of development. Many of the ideas for this book were shaped by discussions with colleagues and mentors and I have been very fortunate to have learned from peers at conferences and symposia. I would also like to thank those who read through drafts of chapters and of the book and offered feedback and advice to me. Finally, there are those in my personal life who have been instrumental in helping me to complete this draft, particularly during great upheavals in my life. I need to thank my family for the endless support they offer and the encouragement they give. And, in particular, I am indebted to James for his patience, proofreads and good humour at all times.

Introduction

This book is primarily concerned with women's relationship to the television industry in its early years and the discourses that framed this relationship. It considers women as audiences, consumers and workers. To account for this relationship – what determined and governed it, what conditions were placed on it, who defined and regulated it – I reach into television's pre-history. Television, as I argue throughout the book, was gendered not only at the point of its emergence as an industry but even prior to this when something *like* television was imagined, proposed, invented and practised. The invention of television and the emergence of a television industry were dependent upon a host of other inventions and industries such as the telephone, photography, film and radio. In this sense, television was a remediation of other technologies and media. Television also reproduced socio-cultural conventions that ultimately determined what role women could play in it, what investment they could have in it and how they would be addressed by it. Such conventions were established in other technology, media and entertainment fields. Technical and engineering work, for example, was male-dominated, with women largely excluded from such professions. Theatre and film industries and cultures facilitated practices of consumption and reception that were sex-segregated – not necessarily to exclude women, but to provide them with a comfortable and safe experience, and thus attract larger audiences. Radio broadcasting, having been initially relatively open to women as broadcasters and producers, quickly reframed women's role as its audience and not its makers.

As the promise of television materialized in the popular press from the nineteenth to the early twentieth century, women's relationship to this new technology was up for negotiation. Thus, at least initially, entry into the television industry was not excessively prohibitive, since television was only considered a side-attraction to radio. Once television professionalized in the post-war years, the same patterns of segregation emerged and women's relationship to television was defined not through their production of it, but their consumption. The sphere of production, associated with technology and commerce, came to be understood as masculine. Television consumption, associated with the 'masses' and passivity, came to be understood as feminine. This binary has largely informed women's role in television, although, as I argue, it has not always been the case in practice.

Part of the purpose of this book is to examine the prospects for, and discourses of, women in screen and broadcast media at their point of novelty and promise from the mid-nineteenth century to the mid-twentieth century. This history is framed by television, beginning with how it materialized as an idea and culminated in an industry and form that continues as one of the largest mass media of our time. Many of the historical accounts of the rise of television – its arrival first as an idea and then potential technology, its invention and its development as a medium – have concentrated more on men's role as inventors and innovators and less on the prospects it offered women, the discourses that materialized about women and television or, indeed, women's stake in emerging television industries and production. Early literature on the development of television accounted for it as a masculine technology in the sense that it concentrated on the achievements of men in the initial efforts to develop a viable method of transmission. Books on television by Charles F. Jenkins (1925),[1] Kenneth A. Hathaway (1933),[2] Philip Kerby (1939)[3] and Robinson Hubbell (1946)[4] detail the history of technological experimentation and innovation, one inevitably defined by its maleness since women were largely excluded from technical and engineering fields during this time.

This practice was equally reflected in the many biographical accounts of television history which detailed the 'great man' stories of how television emerged as the dominant medium, such as Jenkins'

1931 autobiography *The Boyhood of an Inventor*[5] and George Everson's
1949 biography of Philo T. Farnsworth.[6] As later historical accounts of
television expanded to include the emergence of television production
practices, of the structure of the television broadcast industries and the
form of television programmes, they often remained equally focused
on the actions, roles and achievements of men in television, with some
brief acknowledgement of women as some of the faces of television
or as part of the television viewing public.[7] While these histories
are not framed as histories of men and technology, they ultimately
read as such because of the absence of women. In other words, they
are male by default. As Wendy Faulkner suggests, 'technology is
gendered because *key specialist actors* – especially in the design of new
technological artefacts and systems – *are predominantly men*'.[8] Given
the dominance of men (or rather the dominance of attention paid to
men) as innovators, inventors, engineers, business owners, financers
and producers of television, women have consequently been excluded
from or been peripheral to the history of television's emergence as a
medium. According to this technological and economic history of
television's rise, women were invisible.

Women and technology

Feminist histories of technology have sought to address this issue and
to interrogate not only this absence but also the gender/technology
relation that is productive of it.[9] Judy Wajcman, following Cynthia
Cockburn, asserts that the dearth of women in technical work is not
simply a result of rigid barriers of entry but also related to 'the sex-
stereotyped definition of technology as an activity appropriate for men'.[10]
Women are not simply excluded because they don't have access to the
skills and tools, but because the culture and language of technology are
masculine.[11] Therefore, even when some access is afforded to women,
as in the case of early television, the very organizational and business
practices that emerge already work to alienate them. It is not simply the

case that development and production are defined as masculine because men practise it, but that their practise of it 'cultivates relations amongst men'.[12] This results not only in a technical barrier to entry for women, but in a cultural barrier to entry whereby the organization or institution is simply 'not for women'. The consequence of this is that the technologies that emerge from such cultures, organizations and institutions 'bear the imprint of the people and social context in which they developed'.[13] This gendering of technology manifests either directly or indirectly and in a whole set of ways; for example, in terms of how technical and administrative roles in broadcasting were largely sex-segregated,[14] how the pitch of women's voices was considered incompatible with radio broadcast but their faces were deemed suitable for television,[15] or in the concerns broadcasters had about the disruption television would cause to the housewife's domestic work.[16]

This imprint that Wajcman writes of is reflected in Ruth Schwartz Cowan's *More Work for Mother: The Ironies of Household Technology from the Open Hearth to the Microwave*, whereby 'technical progress' that was achieved through mechanization of domestic appliances, rather than reducing the workloads of women in the home, resulted in an overburden of work for women.[17] Cowan's work turns attention away from the production of technology to its consumption. Her account of the introduction of technology to the home helps us to understand how the domestic space was encroached upon by technology and how it contributed to the redefining of gender roles in ways that inhibited women. While Cowan concentrates on housekeeping and household technologies such as the stove and the refrigerator, it is possible to extend this to television since similar processes of gendering took place in relation to women's use of the television. Indeed, the gendering of roles via the use and consumption of domestic technologies persisted in the era of television whereby producers of television cultivated the notion of the female daytime audience, thereby encouraging the separation of domestic activities by gender. In effect then, each stage of the development and practice of television – from innovation, invention, production practice, to its use and consumption – was embedded in what Maria Lohan and Wendy Faulkner term a 'gender-technology relation'.[18]

Women and consumerism

Indeed, a focus on consumption and consumerism as part of the process of technological development opens up a route to television that is inclusive of women since it was primarily in this capacity that women were associated and engaged with television in its early years, and as one group of stakeholders invested in the promise of the emerging technology and medium. However, women's role as consumers cannot be extrapolated from the myriad of competing meanings that have been attributed to consumerism more generally and the interrelationship between gender and consumption more specifically. Feminist historians of consumer culture and consumerism have, as Victoria de Grazia notes, taken a range of (sometimes competing) approaches to, and developed a multitude of perspectives on, the relationship between women and consumer culture.[19] For some, the rise of consumer society and commercial culture in the post-industrial era extended the oppression women experienced in Western capitalist patriarchy. This is perhaps what Rita Felski means when she suggests that the 'feminization of modernity ... is largely synonymous with its demonization'.[20] On the other hand, as de Grazia suggests, feminists have argued 'that mass consumption liberates women by freeing them from the constraints of domesticity' whereby women can 'use the rituals of consumption ... to bend the norms ordained by the market and to flout family and other authority'.[21]

Jennifer Scanlon echoes this when she writes that '[a] good deal of scholarship has focused on the disempowering elements of consumer culture: the ways that people are encouraged to buy beyond their means, females are portrayed as either simple-minded consumers or the bodies that provide male consumers with attractive consumer accessories'.[22] She also points to the body of literature that finds 'that consumer culture can be liberating' in the sense that consumer culture can undermine, disrupt or offer pleasures to women not afforded within culture more generally.[23] This effort to recover women's consumption has, in turn, been undertaken by scholars of popular culture such as in Janice Radway's *Reading the Romance: Women, Patriarchy, and Popular Literature*,[24] Miriam Hansen's *Babel and Babylon: Spectatorship in American Silent Film*,[25] Shelley Stamp's *Movie-Struck Girls: Women and Motion Picture Culture after the Nickelodeon*[26] and Ien Ang's *Watching Dallas: Soap Opera and the Melodramatic Imagination*.[27] Feminist television scholars have considered these complex perspectives in respect of consumer culture and women's relationship with television, particularly regarding the gendered address of television, the domestic location of the television set and the role and representation of women on television screens.[28]

Among those criticisms of consumer culture were claims that it inhibited the social role of women and resulted in a derogatory definition of mass culture as feminine. Andreas Huyssen, for example, noted that it was 'striking to observe how the political, psychological, and aesthetic discourse around the turn of the century consistently and obsessively genders mass culture and the masses as feminine'.[29] Ann Douglas likewise argues that the emergence of mass culture was accompanied by, if not dependent upon, the relegation of women from producers of culture to the unproductive role of consumer.[30] In contrast, others find some form of female liberation possible in the emerging consumer economy, particularly in terms of women's practices of consumption, leisure, recreation and shopping. Some found that the rise of public spaces of consumption in Britain, such as the department store or the shop window, allowed middle-class women to engage in the activity of shopping which, in turn, 'enabled a variety of cultural and discursive constructions' of identity.[31] Others have argued that public sites of consumption played an important role in affording women agency as consumers.[32] In particular, the power afforded to the act of looking and observing challenged some of the ways that consumer culture reinforced traditional feminine domesticity.[33] Ultimately, women's access to and experience of commercial and public forms of entertainment such as theatre, amusements and cinema were bound up in gender politics that struggled to synthesize the idealized image of femininity with women's material practices.

Women and television

Women's relationship to television technology and television consumption, then, was subject to regulation and management as was the case with other media that came before. Women were excluded from participating in the invention and development of television. In addition, their role as consumers was largely defined for rather than by them. They had more access to the sphere of consumption, but this access was contingent upon a number of gendered discourses often relating to their role as wives, mothers and carers. This trend continued with the introduction of radio and television, albeit with a number of notable differences. Women in radio and television were eventually excluded from most senior or authoritative roles, but not after quite a number of them had already gained access to the industries during their infancy. Those recuperating women's contributions to television continue to map the roles, experiences and productions of women in both British and US television. Scholars have uncovered significant contributions made by women, particularly during the 1940s and 1950s during which women's participation in television was structured by a variety of social, economic and ideological conditions.[34] However, the popular narrative of radio and television production as the terrain of men persists and women's contributions have yet to be fully accounted for. The aim of this book is to contribute to feminist scholarship of women in television's pre-history and early years and to generate a narrative about women's relationship to emerging media and entertainment in the years leading up to television.

More attention has been paid to female consumers of radio and television, particularly in regards to women's reception of radio and television from the 1950s onwards as well as the way that television shaped the domestic sphere.[35] However, less attention has been paid to how the female audience was produced by the radio and television industries and how this impacted the forms that radio and television took. Some scholars have offered more general critiques of institutional audience research. Mark Balnaves, Tom O'Regan and Ben Goldsmith point to weak data collection methods that result in incomplete pictures of the audience.[36] Ien Ang argues that the institutional concept of the audience is essentially a fiction that masks the real practices of the actual viewers and listeners.[37] Dallas Smythe proposed the concept of the 'audience commodity' as a way of explaining how audiences were the product formed and exchanged within media industries as well as how listeners and viewers engaged in forms of labour.[38] Eileen Meehan instead argued the 'commodity audience' more aptly described the situation in which prices were set for particular audiences.[39] For Meehan, audience measurement and ratings categorized audiences and applied these values. These audience values were shaped by the ideological and patriarchal norms of those in the media industries and this resulted in the devaluing of the female audience. While Meehan refers to the use of ratings more generally, my book addresses how qualitative audience research worked to produce discourses of the female audience. Indeed, while studies of reception point to significant differences between the institutionally produced concept of the audience and the material practices of listeners and viewers, few in-depth studies of audience research reports and studies have been made.

Gender and discourse

Throughout this book I refer to and analyse discourses that emerge from texts such as newspapers, technical journals, magazines, promotional material, press reports, audience surveys and internal corporate and company memoranda. Discourse analysis is used as a way of examining a range of ideas, perspectives, stereotypes and views of women and their relationship to television that are found in such texts. I examine the language used about and the attitude to women as expressed across such texts during the nineteenth and early twentieth centuries. The image of woman that was discursively produced and, in many cases, normalized and idealized was middle-class, white, domestic and maternal. Thus, the women referred to throughout this book are, for the most part, those who fit this narrow mould. Television and television discourses, in this respect, perpetuated 'hegemonic racial' hierarchies and structures as well as those of class and gender.[40] In the press, marketing and trade discourses examined throughout this book, womanhood was closely tied to middle-class identities, to youth, often to motherhood and, more particularly, to whiteness. This was evident in the institutional address to and understanding of women[41] and in the industry's promotion of 'television girls' – those women who worked in front of or behind the television camera in the formative years of the industry and who fit racial, class and heteronormative ideals. In the BBC, for example, women's participation in television production and roles as television announcers were acceptable because such women were largely educated, white and middle-class.

Therefore, while women – as compared with men – were underrepresented in television production, on television screens and in television discourse more generally, there were levels of underrepresentation across women. The women who had more access to and opportunities in television and those who were more often institutionally addressed tended to be those who belonged to the same class and racial group. Institutional and popular discourses about women and television ultimately perpetuated social hierarchies and power structures. Ultimately, then, this book often has as its subjects middle-class, educated and white women from Britain and the United States. There are, of course, many other stories to tell about women in early television.

While this book is mainly concerned with women and early television, it also investigates how gender discourses emerged from the nineteenth century and how such discourses laid the groundwork for women's later participation in television. Although these discourses – of women's role as representatives and representations, as workers or as audience – did not necessarily determine the conditions of women's access to and participation in television, they suggested certain gendered norms that would structure television. For example, where there was some hostility towards female radio announcers, women were largely accepted and celebrated as the faces of television. Women could be the object of the television gaze but not the speaking subject of radio.[42] Following Foucault, who suggested that people are categorized and regulated through discourse, I examine how power hierarchies were exercised and negotiated through media discourses on women's access to, use of and participation in television and other media.[43] As such, I consider discourses as constitutive of, as well as reflective of, social relations.[44]

My analysis demonstrates how women were produced and addressed as a group invested in television and other media, such as cinema and radio. Women, this analysis suggests, were often present in press features and articles about emerging media and were imagined to have certain investments in them. These discourses were meaningful in the sense that they preceded the actual roll-out of television and provided ways of understanding women's relationship to media and television. These meanings were dependent upon the intended audience for each text. Hobbyist magazines of the 1920s and 1930s, for example, perpetuated gender stereotypes of women's technical incompetence since these magazines were aimed at male enthusiasts.[45] Trade magazines and industry-facing literature gendered women's consumption of media including radio and television. My analysis of the discourses that circulated about women and television and other media is concerned with this process of gendering.

Throughout this book, I refer to gender discourses that materialize in and across various media during the nineteenth and twentieth centuries. The term 'gender' is used as a means of describing the cultural, behavioural and performative norms that were assigned to the male and female sexes during the period under discussion. The phrase 'gender discourse' is used to refer to the ways that various media texts collectively reproduce gender norms either implicitly or explicitly. I examine how gender discourses are constructed in the ways that women are referred to, discussed, mentioned and even omitted in articles, papers, stories, advertisements and other texts that are concerned with entertainment and media such as theatre, cinema, radio and television. I refer to women for the most part although I recognize that the women I discuss often share class positions and racial identities – as middle-class and white – and that this particular intersection of gender, class and race forms much of the discussion of this book. In my discussion of gender, I am concerned with how women's relationship to media was gendered in the press and the media. By gendered, I mean the ways that media represents the status of being female in particular ways that are often oppositional to men. I assess the way that media representations of women conform to, perpetuate or produce certain meanings about the female sex that are in line with the social and cultural expectations of being a woman. I view gender, more generally, as a social construct that (at least during the period under discussion) reproduces male dominance and female subservience. In my analysis of representations of women and in my discussions of the media discourses of women, I see such media texts as producing a sense of 'what it means to be a woman'.[46] The media's reporting of women and their relationships, interactions and engagements with television and other media did the work of defining women's place in relation to such media. In this sense, media reports gendered women and aspects of the emerging media industries, including job roles, types of content and audience tastes.

Methodology

This book focuses more on discourses about women rather than the direct experience of women and their engagement with emerging media such as theatre, cinema, radio and television. My concern is with 'ideas about a phenomenon' rather than the phenomenon itself.

I examine these ideas and discourses in order to understand how power relationships were formed, negotiated and debated in relation to new media of the period. The period under discussion is broadly that of the pre-history and early years of television. This includes the period in which a range of other media and entertainment forms emerged including cinema and radio. I begin with an examination of the ways that women's patronage of theatres was discursively produced in the media during the 1850s. I examine the early references to 'television-like' technologies in the 1890s and early 1900s. At the latter end of this history, I discuss the trade press' reporting of women's exit from Chicago television station WBKB in 1946. I end with an examination of institutional audience research such as that of the BBC in the early post-war period. Although these are separate eras and separate media, I connect them through the variable of women and media discourse. I argue that before broadcast television existed as an industry and a form, it was under development not only as a technology but also as an experience. The technology, the form and the culture of television drew much from preceding media. I make the case that the relationship that women were imagined to have with television was shaped by discourses of women's role in theatre, in cinema and in radio. Equally, I suggest that the ideas that circulated in the popular and trade press about women and television prior to television's arrival demonstrate the extent to which television was gendered before it quite existed.

Because this book focuses on the gender discourses evident in the popular and trade presses, it is limited in what it can say about the history of women and television. In other words, this book is not a biography of various women in the television industry, nor does it offer an exhaustive account of every encounter of women with early television. Instead, it examines those women who have not necessarily become part of dominant television history but who were referenced in and profiled in the popular and trade press of the day. It considers women across a range of roles from directors and producers to announcers, craft and technical crew. Of particular interest are those women who emerged from the experimental years of television: Mary Adams and Mary Allan of the BBC; and the Women's Auxiliary Television Technical Staff of Chicago experimental television station WBKB. It looks at the 'television girls' who acted as promoters of television before its launch including Betty Goodwin and Natalie Towers. Collectively, they evidence the participation of women in early television beyond the famous pioneers. The selection of women discussed is by no means a holistic picture of women's participation in television production, and there are many other cases of women working in experimental, and commercial and public service television outside of those referenced. However, I hope to provide a picture of what discourses materialized about women in television production through these case studies.

This book also examines the emergence of gendered discourses of television consumption and the television audience. The resources used to make a case about gender discourses include: newspapers such as the *Times* and the *New York Sun*, trade publications such as *Billboard* and *Sponsor*, entertainment and broadcasting news publications such as the *Radio Times* and *Variety*, technical and/or hobbyist magazines and publications such as *Short Wave & Television* and *Wireless World*, corporate and/or internal publications such as *NBC Transmitter* and the *Listener*, inventors' publications such as *Vision by Radio: Radio Photographs*[47] and *The Victory of Television*,[48] and audience studies and surveys such as the BBC's report 'Viewers and the Television Service'[49] and Advertest's study 'The Television Audience of Today'.[50] I use these sources not as confirmed first-hand accounts of the status or role of women in television and other media, but as texts which can be subject to discourse analysis. These publications were addressed to different audiences in different contexts and the gender discourse(s) that emerged from, for example, entertainment news publications that were addressed to the public and trade publications that were addressed to broadcasters and businesspeople are telling. Throughout the book, I situate these in their socio-historical and economic context. For example, I note that advertising trade publications were concerned with a specific idea of woman as white, middle-class and married. Such publications imaged women's role as consumers of television. This discourse of the female audience said less about the actual activities, relationships and behaviours of women regarding television than they did about a commercial and capitalist ideology that had reimagined gender in distinctly consumerist terms.

In contrast to the aforementioned sources are those materials, documents and artefacts that speak to the institutional or corporate practices, the national policies, the personal experiences and perceptions of those involved in the emerging media industries. While these are referenced insofar as they provide context to the discourses that emerged in the popular and trade press, they are supplementary to my analysis of gender discourses. For example, I refer to internal memos and correspondence regarding the treatment of BBC television programme-maker Mary Adams in order to compare her treatment by BBC senior management with the promotion of her in the press as representative of the BBC's gender equality. This undoubtedly risks silencing women's own voices and experiences and in framing women as a discursive category rather than as embodied people with agency. However, the place of women in television's pre-history is difficult to identify otherwise. In the case of television's invention, for example, records are kept of the inventors, of patents and of business. The names recorded on these tend to be male even if some of those involved were female. As Rebecca Ann Lind notes, 'the social status of women … [has] affected the extent to which their contributions are celebrated, preserved or even acknowledged'.[51] The technical, engineering and business documentation relating to early experiments with television says very little about whether women were involved and what their contribution was. When these same experiments took place for the press, we find more reference to women in attendance, to their interest in it as viewers and to their roles in the transmission of images. Female journalists such as Violet Hodgson (who witnessed a ship-to-shore television demonstration in 1931) recorded their feelings about television and imagined an egalitarian medium that would benefit humanity.[52]

Yet, in most cases, women did not speak (through these publications) but were spoken of. What was said and how women were represented indicate, to some extent, the tensions that were raised regarding women's role in emerging media. Inasmuch as the popular and trade press evidence acceptance of women in spaces such as theatres, as workers in radio and television or as audiences, this was conditional acceptance. The discourses within which women were framed did not necessarily determine women's actual experiences of television but to a certain degree gendered them. My objective, therefore, is not to infer that mere reference to women was enough to evidence what women actually made of television or what their role in television was. This book does not provide an exhaustive account of actual women and their relationship to television and other media. Instead it assesses the discursive strategies deployed in the popular and trade press to suggest the terms of women's engagement with television and other media.

Scope of the book

This book, then, asks the following questions of women's relationship with television: How did the invention of television technology work to shape women's access to and experience of it? What was women's experience of preceding media and entertainment, and how did this inform their later access to television? How was women's access to media and entertainment regulated and structured, and how did this influence their access to television? How did women participate and work in early television? And how were women conceptualized as the female audience? In answering these questions, I look to the United States and Britain as case studies. Among the reasons for this are the fact that the United States and Britain were early pioneers of television technology and were among the first nations to develop television industries. In both cases, experimental television had commenced by the 1930s with women partaking as hosts, announcers, producers and technical workers. These early histories are often neglected outside of a few 'top-down' histories that concentrate on regulation, policy and industry leaders. Histories of women in US and British television often begin in the post-war years, when full broadcasting systems were developed and television set ownership had increased dramatically. These histories have begun when television was 'established' (from the late-1950s onwards). Much less attention has been paid to the preceding decades when, even without a full broadcast service, television was very much in the popular imagination in both the United States and Britain.

In addition, a comparison between US and British television reveals differences between both industries at the level of structure, regulation and form. US television was commercial where British television was a public service. However, there are some interesting similarities in women's relationship to television in their respective nations. In other words, despite some key differences in how women accessed television or were addressed by it, there were some common discourses. In both the United States and Britain, women were more readily accepted as the faces of television than they were as the voices of radio. In addition, US and British women gained access to television production during the war years but were marginalized following it. Equally in both cases, institutional audience research worked to produce an ideal image of the female audience that spoke to feminine, domestic ideals that had been challenged by women's entry into the workforce during the war years. These shared experiences enable me to construct a chronology of sorts and to identify moments of convergence and deviation in the history of women in early US and British television.

It is an incomplete history as it does not address early and experimental television in other nations such as Germany. Nor does it consider how women fared in nations in which television emerged in later decades. The reader will also note that I have refrained from discussing those female broadcasters whose work is widely recognized and accounted for: such as Hilda Matheson, Gertrude Berg, Mary Margaret McBride and Lucille Ball. These histories have been well-mapped by scholars such as to Michele Hilmes, Donna Halper, Susan Ware and Cary O'Dell. In addition, I hope to emphasize the presence of women beyond the oft-referenced and aforementioned pioneers. Indeed, my initial premise was that these broadcasters laid the groundwork for other women to pursue television careers. I came to find that there were many other women working in television prior to and during the time of these pioneers. I am also cautious to avoid perpetuating an image of exceptionalism, whereby individual women are heralded as unique in their television work. Ultimately, in offering some stories of women's place in the very early history of US and British television, I hope that it goes some way towards filling gaps in the history of women in television.

Chapter 1, 'Nineteenth- and early twentieth-century gender and technology', traces the pre-history of television in order to demonstrate how it was a gendered technology even prior to its material reality as an institution. I discuss how technological innovation of communications and media technologies in the nineteenth century was largely the sphere of men. Understanding this history as a history of the work on men allows us to understand how it was inevitably gendered. I examine the role the press, as communicators of such inventions, played in gendering the social use of such technologies. I examine the early public displays and exhibitions of media technologies, which, like the salacious mutoscopes that showed indecent images thought to corrupt men, often suggested gender consumption of them. Equally, women's access to public spaces and venues had to be carefully negotiated so they could be accommodated. On the one hand, this helped women access more spaces, but, on the other, it meant that women's access was dependent upon conforming to gendered norms regarding what was appropriate behaviour. Magazines, newspapers and theatre establishments worked to set these conditions. In this chapter, I also discuss how women's role as consumers helped justify and normalize their participation in and engagement with media entertainment. While women were invited to partake in public consumption, they were consequently trivialized and demeaned, as was the case with 'the matinee girl' whose film fandom was considered childish and frivolous.

In Chapter 2, 'Television's earliest years', I discuss the development of television during the early twentieth century and up until the late 1930s when the Second World War interrupted its expansion. Television had, during this period, become a popular topic, with inventors and the press attempting to determine what television would be, how it would be used and by whom. Television's 'gender' formed part of the debate. As a domestic medium it offered the potential to be used by women, but as a material technology it was considered masculine and, therefore, too technical and complicated for women to make use of. As such, men were imagined as the controllers of the television set in hobby magazines and the press. Equally, women's presence on television was divisive. While women were considered suitable as objects of display, there were many concerns about their roles as hosts or announcers. Such roles were considered too authoritative and prestigious for women to undertake.

The next two chapters – 'Women in early British television' and 'Women in early US television' – discuss how women fared behind the camera. In the case of Britain, I note that the main broadcaster, the BBC, had a more equitable work culture, at least in name. I discuss the work of some women who held key roles in television production during the years of experimental broadcasting from 1936 to 1939, including that of announcers Elizabeth Cowell and Jasmine Bligh; producer Mary Adams; and Head of Costume and Make Up, Mary Allan. However, as the BBC professionalized in the post-war years and television became a 'serious' enterprise, women found themselves increasingly sidelined, through deployment to more feminine roles or feminine productions. Women who did hold senior roles in production were often found in sex-segregated departments such as women's programmes, daytime television or children's and educational programmes. I argue that while these women were individually successful in their fields, the gendering of television resulted in the devaluing of their output. In the case of US broadcast, I find similar patterns to that of British television. For example, I note how women benefitted from the relative openness and informality of early television during the late 1930s and early 1940s. Since early television was a side project for most stations and networks, and not yet structured and professionalized like radio was at this point, women were able to take up roles in station management, in producing and directing as well as in craft and technical roles. In addition, I note how women were recruited into television production and technical work during the war since, unlike Britain, television transmission continued during this period. However, I argue that, like Britain, the post-war professionalization of television made such access more difficult and, once again, women found themselves working in roles that had, by now, been gendered.

The final three chapters turn to the concept of the female audience as it emerged in marketing and media research. In the chapter 'Populations, consumers and audiences', I discuss how population measurement, consumer and market research and radio audience research shaped populations into manageable and knowable groups and categories. I argue that consumer research was more productive of gender than it was observant of it. I examine how consumer research emphasized and exaggerated gender differences, with the result that the female consumer was understood in highly conservative and traditional ways. This produced an identity and profile of the consumer who was attractive to advertisers and salespeople. A range of advertising publications such as *Selling Mrs. Consumer*[53] helped define women as female consumers by associating them with domestic purchasing power, with their roles as housewife and mother and with their naivety and susceptibility to advertising. I then discuss how this idea of the female consumer increasingly informed radio research carried out by advertising-supported broadcasters as well as consumer and social researchers. In the United States, in particular, great effort was given to producing a profile of the female radio listener since she was considered the primary purchaser in the household. This resulted in the perpetual gender-differentiation of the audience that eventually led to gender-differentiation in programmes and schedules.

In the chapter 'The US female television audience', I argue that such consumer and audience research set the conditions under which the US female audience would be understood and acted upon. While early television audience research was less concerned with profiling individual viewers, research from the mid-to-late 1950s increasingly promoted the idea that women watched television in the daytime. This, I suggest, worked to establish the daytime as the terrain of female viewers and the evening schedule as that of men. This was a 'common sense' assumption drawn from radio audience research that was, in the case of television, much more difficult to evidence. As I demonstrate, television audience research, such as that carried out in the Videotown surveys of the 1940s and 1950s, consistently demonstrated that women watched more in the evenings than in the daytime and often more than other family members. In the final chapter 'The British female television audience', I note how audience research arrived later in Britain, and was not integrated with the advertising industry, since the BBC was a public service broadcaster. Nonetheless, a similar pattern emerged, I argue, in which audience researchers increasingly profiled the BBC's viewers by gender. I examine the reports and surveys carried out by the BBC Listener Research Department (later Audience Research Department) in order to identify how and why this occurred and what the implications were for female viewers. In producing the concept of the female audience, particularly in the post-war period, I suggest that BBC audience research worked to re-domesticate women by defining the female audience.

1

Nineteenth- and early twentieth-century gender and technology

The late nineteenth and early twentieth centuries saw transformations in and rapid acceleration of both media and entertainment technologies and gendered social relations. These were, in many cases, interrelated and co-constituted. Early media and entertainment technologies were both reflective and productive of gender discourses and this is evident in the competing discourses that circulated around the emergence and social use of such technologies. Speculative technologies – those that were not ready to be deployed but were popular news topics – lent themselves to fantasies and debates about how they would reinforce existing gender hierarchies, how they might challenge or upset conservative gender norms or, in contrast, how they might form part of women's social liberation. Contradictory narratives emerged in the popular press, promotional literature and even science fiction literature whereby women were represented as beneficiaries of progressive and liberating technologies according to some, and victims of these same technologies according to more conservative accounts. This was reflective of the wider social debates, fears and backlashes against the transformation of women's role in society. Women's access to and engagement with technologies and entertainment experiences such as the mutoscope and the moving picture were scrutinized and debated in newspapers and periodicals of the period. Women's place in this new techno-entertainment culture was, therefore, negotiated and regulated by social gatekeepers, cultural commentators, promoters and businesspeople, and by women themselves. This late nineteenth- and early twentieth-century gender-technology relation is examined throughout this chapter.

The chapter discusses the implications of the gendering of technological invention and development of early screen entertainments. It suggests that such technologies – being designed and developed by men – began as masculine. Women's use of them was therefore not immediately 'obvious'. Instead, women's relationship to media and entertainment technologies was contemplated, contested and debated. What emerged was a range of discourses about women and representation as well as women and consumption that would set the conditions of women's role in television culture when television eventually arrived.

Such discourses must be seen in light of the changing status and role of women in British and US society during the period. Women were engaging more in work outside the home. They were participating in politics and making their voices heard in the political sphere. Women were advocating for suffrage and for education.[1] Women were also becoming important players in the field of consumption and became central to the address of the consumer industries and advertisers. Such advances and achievements were celebrated and championed at times in the press. At the macro-level, at least, women's social position seemed to be improving. However, this was accompanied by, in some cases, scepticism and, in other cases, a backlash. In other words, there was no teleological and holistic advance of women's position during this time.[2] Women who entered the political sphere were subject to criticism, as were those who pursued women's suffrage. Working married women faced a backlash as a new domestic ideal emerged to challenge the shift of women from the home to paid employment.[3] Even in those cases where women were celebrated and praised as pioneers and trailblazers, this was presented in the press as exceptionalism rather than as something that all women could aspire to.[4] For Ann Oakley, narratives of women's domestication and the construction of the housewife identity emerged precisely as a response to industrial capitalism.[5] Victorian discourses of gendered division of labour worked as a counter-point and an antidote to women's increasing economic role as consumers and workers. Women's consumption, industriousness and productivity were valued insofar as they contributed to an ideal domestic life.[6] This was perpetuated in women's magazines and periodicals such as *Good Housekeeping* and *Ladies' Home Journal* that favoured women's domestic life and sanctioned certain forms of womanhood.[7]

In contrast was the popularization of the image of the New Woman who was the focus of much attention for her pursuit of interests and activities beyond the domestic space. The discourses of the New Woman that appeared in magazines and periodicals such as *Punch* evidenced the anxieties about and efforts to make sense of gender and social change that accompanied technological and lifestyle change.[8] These competing ideas of womanhood might be understood, in some ways, as a response to women's encroachment upon typically male spaces and roles. This anxiety about women's increasing presence in and engagement with typically male pursuits is evident in the discursive responses to the development of entertainment and communication technologies and the gendering of them throughout the late nineteenth and early twentieth centuries.

Technology as male

Although television is, according to some scholars, a feminized medium,[9] most of the renowned figures of television's pre-history, at least, are those male engineers, electricians, scientists, entrepreneurs and businessmen who have been identified as the significant characters in the story of the rise of television.[10] Equally, those newspaper reports of experiments and innovations, of public lectures and demonstrations and reviews of latest developments in 'seeing at a distance' – using emerging technologies to send and receive images across space and time – represent largely the male voice. In the pre-history of television, and at a time in which many of these communication technologies were not yet socially prevalent, the discourse of 'seeing at a distance' was predominantly a male one. There are a multitude of reasons for this, perhaps the dominant being the persistent exclusion of women from education and the professions. Thus, where female ingenuity was often commented upon in the press,[11] and while women did participate in technical innovation (as evident through applications for patents, for example),[12] these were largely noted as a novelty in the popular press. Furthermore, such press reports repeatedly made claims about female firsts such as 'the first female engineer' or 'the first female mechanic' even though many precedents had already been set.[13] This both validated women's participation in technological innovation and, at the same time, perpetuated the myth that women did not belong in this field, since the pioneers referenced were figures as the exceptions to the rule.

Judy Wajcman suggests that part of the feminist project has been to query 'whether the problem lies in men's domination of technology, or whether technology itself is inscribed with gender power relations'.[14] Cynthia Cockburn has accounted for this by tracing the ways in which men have actively sought to dominate certain spheres of communication technology.[15] She notes that masculine identity was (and is) defined in relation to technical achievement and skill. As a result, women were largely marginalized within such labour spheres. Thus, the supposedly 'natural' inclination of men to roles in engineering, design and manufacturing reflected more the gendered division of skills and labour than it did real sexual differences among men and women. This produced forms of social power over women and also meant that new technologies were imagined and developed in ways that reflected and constructed gender difference. This monopolization of technology resulted, therefore, in technology being both produced and defined by men, with women addressed as consumers or users following the mainstreaming of such communications technologies, and only then, once advertisers and salespeople had identified a valuable female market. More generally, technological innovation was predominantly carried out and recorded by men. This took place as much in the field of media research, development and innovation as it did elsewhere. Consequently, histories of early experiments with 'seeing at a distance' technology tend to focus on the practices of male inventors. The project at hand, therefore, is to more fully account for the ways in which the types of technologies developed within these years were primarily constituted via 'historically and culturally specific norms' of masculinity.[16] Early 'seeing at a distance' technologies, on the one hand, had a multitude of potential uses, yet, on the other hand, they were guided by masculine interests.

Since television technology was largely the product of male inventors and was the object of male (business and technical) interests, this had implications for women in terms of how they entered the field of television as producers and as consumers. For Cockburn, an analysis of technology as gendered reveals the asymmetries, the aspects of control and domination that are experienced by women who engage or interact with technology as defined, organized and produced by men.[17] During the late nineteenth century, there were many speculations about the social, political and commercial utility of early television technologies. These utilities were proposed by researchers and scientists but also within the wider press and in popular literature which, at this time, evidenced a great deal of interest in such technologies. These ideas that circulated in the press as well as technical and professional journals drew from historical precedence (telegraphy and telephony had already confirmed that it was possible to send messages across vast distances) while also imagining future possibilities. At times, women were included in this technological future. At times, they were not.

A distinctly masculine perspective of television technology was, for example, evident in a paper published by Adriano de Paiva for *O Instituto* in March 1880 titled 'A telephonia, a telegraphia e a telescopia' (telephony, telegraphy and telescopy).[18] De Paiva's paper, which was hugely enthusiastic and excited about the potential of 'seeing at a distance' technology, imagined it as a form of masculine dominance over nature. In using masculine terms and pronouns to describe the mastery of television technology, de Paiva implied a binary between masculine ingenuity and scientific endeavour and nature, which, in the same analogy, was inferred as female.

> When we mediate a little on the way in which communication *between man and surrounding nature* is established, two of the organs of our sense appeal to reveal a superior importance. These are the eye and the ear … it is not surprising that ever since he began to use them, *man should have endeavoured to increase their sphere artificially.* [With new technologies] man will be able to extend to the whole of [the globe], his visual and auditory faculties.[19]

This echoed long-held and equally disputed conceptions of woman as representative of nature, and which man (through science, logic and rationality) overcomes and dominates.[20] The gendered discourse evidenced in de Paiva's article held television technology as an exclusively male pursuit. De Paiva used masculine pronouns when referencing power and dominance through technology. In other words, de Paiva imagined technology not as an egalitarian technology that was accessible to all, but as a technology of masculine empowerment. This 'top-down' vision of a form of television sat in contrast to the 'bottom-up' or more democratic vision that emerged elsewhere in which technologies were not associated with domination and mastery but were imagined as more egalitarian and more inclusive of women.

Women and technology

Competing perspectives on television's social role and of women's relationship to the technology were found in the popular press. In her account of the 'electric dreams' of the Victorians, Verity Hunt comments upon the tendency of the press to prematurely announce such inventions' success. However, she also notes that the public imagination, particularly speculative literature, worked to socially contextualize such inventions.[21] In this literature, illustrated renderings of the technology worked to offer a vision of how the new technology would be socialized. For example, George du Maurier's image of Edison's telephonoscope in December of 1878 is often credited with predicting a version of television.[22] Indeed, George and May Shiers, in *Early Television: A Bibliographic Guide to*

1940, note that this image encouraged other inventors to think about how 'seeing at a distance' might be realized.[23] In du Maurier's imagined world of the telephonoscope, women feature prominently. In the cartoon (Figure 1), published in *Punch*, a mother and father sit by the fireside viewing a live transmission from the Antipodes, where their daughter converses with the father from a tennis lawn.

The placing of the image above the fireplace and in the living room evokes the later placement of the television in the home. Of interest here is the framing of gender in this cartoon. The man and woman act as viewers of the screen; yet, it is the father who speaks in both the illustration and the accompanying text. The daughter, to the left of the screen, also speaks to her father in return, but only to answer his query about the 'charming young lady' who he has been watching. While father and daughter might interact, the other women in the drawing are silenced: the mother is somewhat dismissed by the father where he whispers to his daughter and the 'charming young lady' is subject to the objectification of the father. Here, the domestic integration of technology was already infused with a sense of power

Figure 1 George du Maurier's vision of the application of Edison's telephonoscope. London, 9 December 1878, with permission of World History Archive/TopFoto.

relations and gender hierarchies, some of which would come to be realized in television broadcast in later years.

Elsewhere, women were imagined to have a more active role in the emerging technoculture. The much commented-upon satirical writings and illustrations of Albert Robida, published in his *Le Vingtième Siècle: la vie électrique* (*The Twentieth Century: The Electric Life*) in 1883, speculated about the various developments in point-to-point communication and moving image technology as well as the participation of women in it.[24] These science fiction illustrations already imagined both utopian and dystopian future social applications of moving image technologies for both domestic and public use (in the case of the book's story, Paris 1952). As much as technology enhanced and expanded the spheres of communication, transportation and the fields of leisure and work, they also, in Robida's future, inhibited it. The telephonoscope (a key utility in Robida's imagined future) was represented as a form of interactive communication (videophone) and also as entertainment (to watch theatre performances). Often it was situated within the living room or parlour and was made up of a large projecting screen, accompanied by a telephone situated next to it. People could use the telephonoscope in the home setting in order to 'keep in touch with loved ones'. The telephonoscope also operated as a newsreel, where families watched news events as a form of visual newspaper. In another illustration, Robida represented a form of home theatre, whereby a gentleman sat before a large screen on which he watched a dance performance by a woman. Elsewhere a woman watched a lecture from her home, while another illustration showed a woman watching a live theatre performance from her bed. Robida hinted at the voyeuristic nature of the telephonoscope: one illustration had a collection of men watching a woman sitting on a bed removing her stockings.

Nonetheless, for Robida, women were key users of technologies and gained pleasure and entertainment from them. Women's use of technology was not trivialized (as it would often be later in the case of television). They had 'serious' tastes in the arts. And while women were also imagined as the 'objects of consumption' on Robida's screen media, his future technological world was one in which women were, in some

Figure 2 Pursuing a course of study by 'telephonoscope'. Robida's idea predating Open University by seventy-odd years. Dated 1883, with permission of World History Archive/TopFoto.

respects, emancipated (see Figure 2). Robida's speculative technoculture was a much more inclusive one than de Paiva's. While both were only imagining what television technologies might look like, they did provide the social framing of future technologies that formed part of the effort to determine what television could be, what it would be useful for, and for whom it would be useful.

Women and representation on film

Speculative technologies might have been the terrain of scientists and science fiction writers. However, a number of exciting new technologies were being realized and made publicly available. A range of moving image inventions had not only been demonstrated to the public during the late nineteenth century, but some had also been commercialized. In 1880 Eadweard Muybridge demonstrated his zoopraxiscope: a device that gave the illusion of moving image.[25] The Edison Manufacturing Company invented a peep show moving image device called the kinetoscope in 1891 and which was made commercially available to the public in 1894.[26] The American Mutoscope Company was formed in 1895 and released the mutoscope (a version of the kinetoscope) in 1897.[27] The early development and institutionalization of film may have contributed to the formative ideas about what television as an institution and social entity might be and what women's relationship to later television might look like. Film, unlike television, was relatively more quickly institutionalized and a commercial entertainment industry emerged in the late nineteenth and early twentieth centuries. During this time, moving image technologies were socially constructed by inventors, businessmen and the public. Various stakeholders, social actors and invested groups gave meaning to moving image technologies. Markets were imagined and proposed to the public through the press. The market for moving image devices came to be represented as a gendered one.

In the case of the kinetoscope, the mutoscope and Edison's promised (but never quite delivered) kinetophone, women were represented in the press not as the viewers of moving image but as the objects viewed on screen. In the case of the kinetoscope the default consumer was male and much of the content developed was addressed to this male spectator. Women's role was not as audience but as image to be consumed. This is evident in the early kinetoscope films from Edison. He assigned his employee William Dickson and a team of assistants the task of producing content that would reflect the capabilities of his device. Early films, therefore, and like Muybridge's, showed the body in various forms of motion: blacksmiths at work, strongmen, dancers and boxers. And once he began to manufacture the kinetoscope for lease and sale, Edison 'transitioned from experimentation to production', thus forming a business centred around leisure and entertainment.[28]

The kinetoscopes were leased to syndicates which placed them in parlours for the public to view. Edison's company produced the films that would be distributed to these parlours. The kinetoscope suffered from a significant drawback; it could only be viewed by one person at a time. This meant that, regardless of its popularity, the audience was inherently restricted. And while the parlours in which the kinetoscopes were situated were hetero-social, the first films produced were by men and largely represented masculine interests.[29]

The films displayed in the nickel-in-the-slot machines worked to alienate a mixed gender audience by way of their over-emphasis on masculine forms of pleasure. The emphasis on male sports and female dance and burlesque performances demonstrated the extent to which 'the filmmakers tended to imagine an audience consisting of men like themselves'.[30] Although, according to Charles Musser, Edison's team came to recognize the importance of broad appeal and later expanded their content in order to cater to female and youth audiences, there remained a male bias within the films displayed. The upshot of this was that women now had access to 'the homosocial world from which they had been either excluded or kept at the periphery'.[31] The kinetoscope at once gendered the content and the audience, while at the same time organizing the site of exhibition as one in which gender boundaries could be undermined. The availability of these moving image devices in public spaces did cause some concerns about gender and public morality,[32] particularly with the popularization of the mutoscope, a device similar to the kinetoscope that allowed the viewer to hand-crank and therefore determine the speed of the filmed performance. The main concerns rested on the corrupting influence of erotic images on men. Some were concerned that women might encounter the indecent behaviour of men. In this sense, the peep shows were a masculine activity that would offend women.

Given that the kinetoscope and mutoscope allowed for private viewing in public space, concerns were raised about the type of viewing that was taking place.[33] The mutoscope (or 'what the butler saw' device) offered sexually taboo images of women in erotic poses and performing dances to largely male audiences. Morality groups and offended members of the public raised concern that the content was in poor taste and offensive. Calls for censorship began with a number of cases being taken against the offending companies and distributors. In March 1899, for example, the newspaper the *San Francisco Call* – a crusader against the immorality of commercial entertainment – featured an extended story on the rise of indecent 'sounds and pictures' in the parlours of the city. The feature referred to the 'For Men Only' sign accompanying a mutoscope and claimed to be concerned for those decent men and women who were subjected to such offensive displays. Most objectionable to the reporter was the sheer accessibility of such sites and exhibitions, which were on the main streets and within sight of the general public. The report went on to name the offending proprietors and called for an end to the companies that profited from such 'disgusting' material. The accompanying illustration showed three scenes: the outside of a parlour where men and women passed by unaware of the indecency inside; two boys being corrupted by the mutoscope film that was advertised as 'For Men Only;' and a number of men crowded around a mutoscope eagerly with a number of women and a girl standing in the background, assumedly at risk of corruption by mere fact of proximity. These peep shows were, therefore, addressed to a male audience and discussed in the press as such. The same newspaper had much to celebrate when its self-confessed crusade against the indecent moving images represented on kinetoscopes and mutoscopes resulted in the prosecution of parlour owners and exhibitors charged with showing 'grossly immoral and suggestive' images.[34]

Women had, therefore, a contradictory relationship to commercially available moving image devices. Women were, in one sense, the cause of the aforementioned moral offence since they were the 'indecent material' displayed in the controversial films.[35] However, they were also considered the most likely to be shocked and offended by the shows. In August 1900, for example, the *Standard* (UK) reported on the seemingly extensive incidents of indecent exhibition and display at the Paris Exhibition. Referring to a 'mania for pictorial postcards' of 'a very objectionable character', the report stated that police inspections were carried out on mutoscope shows, resulting in 'orders given for the suppression of the display of objectionable views'.[36] Police inspections continued in Europe and the United States. In December 1900, the *Illustrated Police News* reported that the South Coast Mutoscope Company were exhibiting indecent images of women.[37] The Ladies' Column of the *Evening Telegraph* echoed the sentiments of the *San Francisco Call* when, in June 1901, the reporter called for more protest against the proliferation of indecent material exhibited through the mutoscope.[38] The columnist drew attention to the public venue of the peephole machines and claimed that she had personally encountered such instances of 'bad taste'.[39] Her objection was, in some ways, as much about the gendering of space as it was about the mutoscope's content. While the content was, by all accounts, morally objectionable, the mutoscope's presence in public space also alienated women. This was an issue since women wished to access the same spaces and the same entertainment as men, and the presence of the objectionable mutoscope threatened them. The social spaces produced by the commercialization of leisure and the commoditization of public terrains were, therefore, contested. While the nineteenth century had seen a relative democratization of urban spaces through the formation of public spaces of consumption – arcades, department stores and amusement parks – there was nonetheless a prevalent anxiety about the moral and cultural consequences of this. The more forms of leisure made available and the more access the wider public had to these, the more that calls for regulation, management and censorship occurred.

These concerns about the effects of early moving image technologies on men and women reflect the ways in which the technologies were gendered from their earliest days. Since men were the early inventors and entrepreneurs, it is perhaps no surprise that such men imagined the technology and the commercial use of it through their gendered perspective. The emergence of erotic images of women in early moving images is the most evident example of how such technologies were developed as masculine. However, such technologies emerged at a time when more and more women were participating in public life and entering into public spaces previously restricted to them. Women were eager to participate in the new entertainment experiences offered. In addition, entrepreneurs and businessmen were eager to capitalize on new audiences and consumers. This emphasis on the female audience would become even more pronounced in the broadcasting era.

Women as audiences

As many have noted, the diversification of certain forms of leisure was accelerated by the changing role of women in public and social life.[40] Where early nineteenth-century commercial leisure was constructed around and offered to males, later in the century the commodification and commercialization of public space afforded women more opportunity to participate in public life and entertainment. More particularly, white, middle-class women became a focus of attention for leisure and entertainment businesses keen to exploit these women's increasing entry into public spaces. This new public and cultural sphere did not emerge uncontested and concerns about appropriateness, family values and decency often accompanied new forms of entertainment that were marketed to mixed audiences. During the mid-to-late nineteenth century, for example, middle-class women negotiated their route to public life through the organization of, and participation in, women's clubs, temperance movements and elite educational endeavours such as public exhibitions.[41] These activities and organizations maintained the class stratification that identified the working classes with crass and lowly forms of entertainment and the middle classes with more conservative or elite pursuits. Middle-class women, therefore, could manage their entry to public spaces by foregrounding morality, decency, intellectualism and charity. Businessmen and entrepreneurs were quick to capitalize on this and to direct their address to this audience.[42]

In museums, public lectures and sideshows, the supposedly educational and enlightening tone of the spectacles and displays were foregrounded. Certain time slots during the day were reserved for women and children and events were specifically tailored to suit their interests. Advertisements for shows indicated that they were suitable for, if not exclusively addressed to, women. Exhibitions and lectures were often delivered or hosted by women. Anatomical museums in London, for example, had by the 1850s allowed women to attend unaccompanied and provided female attendants to the visiting women.[43] P.T. Barnum, the founder and operator of the American Museum in New York city, among others, made a concerted effort to welcome middle-class women. From the opening of his museum in the 1840s, Barnum framed his exhibitions and lectures as more tasteful than the many other amusements in the locality. He assured women that his museum would be a safe and morally upstanding venue that would contain no objectionable material. For example, a promotion from 1849 emphasized the integrity of his establishment:

> The Manager pledges himself that no profane word or vulgar gesture is ever introduced here, and that nothing is ever seen or heard which could be objected to by the moral and religious portion of the community ... Such regulations are established and enforced as render it perfectly safe and pleasant for Ladies and Children to visit the Museum in the day time, though unaccompanied by Gentlemen.[44]

The promotion of the museum as a place appropriate for lone women and free of obscenities secured its reputation for middle-class women who were assured of its decency. Barnum, in turn, was assured of a larger audience. In addition, Barnum was vocal in the temperance movement and also positioned himself as family (and, therefore, female) friendly, holding events such as 'baby contests' to encourage female attendance.[45]

This was not without criticism, however, and many (including women) considered Barnum's promotional activities and events as exploitative. Women's magazine *Godey's Lady's Book*, for example, referred to the baby contests as such:

What we would reprobate in the strongest terms is the playing upon the natural and holy feeling of maternal pride simply and solely for mercenary views in the getter-up of the exhibitions. It is to us an inexplicable social enigma that so many mothers holding respectable positions, and some of them positions of influence, should be found ready, under any circumstances, to submit themselves to the degradation to which exhibitors and exhibited are exposed.[46]

Women were, it seems, not entirely convinced that their entry into public space was being managed effectively or honourably. Some recognized that Barnum was merely capitalizing on their enthusiasm for leisure pursuits and that his address to them was opportunistic. This was not an isolated criticism and Barnum continued to defend his museum against others who saw it as lowering the cultural and moral tone of the urban centre.[47] A key theme in the criticisms of the American Museum was the place of women in it. The British satirical magazine *Punch*, for example, published an illustration in which Barnum's representative of female purity, Jenny Lind, was fawned over by a hoard of men.[48] This undermined the promotion of Lind as a symbol of virtue and a point of female, middle-class identification. It also held in suspicion the ways in which audiences and spectators made sense of such hetero-social spaces. For some, women may have had increasing access to public spaces, but they were consequently put into contact with the more vulgar aspects of social life.

Women's magazines and journals such as *Ladies' Home Journal* and *Godey's Lady's Book* attempted to help women negotiate their ventures into public space and advised on what was appropriate or not for women to engage with. *Ladies' Home Journal*, for example, outlined the risks of mixed spaces on the character and reputation of young girls who, if not cautious, might find their reputation 'cheapened' by their presence in public spaces:

Don't encourage young men to call upon you who frequent liquor saloons, billiard parlors or pool rooms. Don't notice men who stare at you on the street, even if it is a well-bred stare ... Don't stand on street corners talking to young men, though they are acquaintances ... Don't accept promiscuous invitations. It only cheapens you, and may draw you into a circle of acquaintances you will regret having formed.[49]

Although women's access to public venues was becoming more acceptable, there was an accompanied anxiety about the meaning and implications of this, particularly in regards to the impact on domestic life. For parallel to the emergence of new entertainment and leisure culture was the emergence of a discourse of separate spheres that reflected the nineteenth-century cult of domesticity that developed around and reinforced middle-class identities.[50] The criticism levelled at women who eagerly partook in entertainment culture evidenced an idealization of more morally conservative, pious and traditional forms of womanhood and cautioned about the dire consequences of women straying into traditionally male spheres.

However, there were spaces that negated distinctions between the separate spheres of public and private. In the department store, the domestic and feminine were interconnected with public space. The department stores and arcades that proliferated in the nineteenth century offered consumption as the means by which women could participate in social life outside of the domestic sphere. Spaces of consumption were framed and promoted as safe, comfortable and proper places where women could be in public unescorted.[51] The boundaries of the department store and arcade with their sanctioned displays of women's bodies and fashions separated those women who entered from other 'street women' outside the parameters of commercial space, such as prostitutes whose bodily displays were illegitimate and unsanctioned. Those entrepreneurs eager to capitalize on opportunities afforded by women's access to public space promoted consumption of fashion, home goods, beauty products and technologies such as the telegraph and the radio to women who equally accepted the opportunity to move beyond the confines of the home and to express some pleasures in leisure activities.[52] The expression of these powers and pleasures was the source of much debate, and an emergent discourse of 'feminization' formed in relation to the perceived association between woman and popular culture: one that would extend into the television era.

That women were participating in cultural activity through leisure, consumption and entertainment resulted in an analogy between femininity and superficiality. Andreas Huyssen, for example, notes 'how the political, psychological, and aesthetic discourse around the turn of the century consistently and obsessively genders mass culture and the masses as feminine, while high culture, whether traditional or modern, clearly remains the privileged realm of male activities'.[53] In other words, a correlation was made between women's participation in social life and mass culture. Mass culture was considered vulgar, simple and trivial. This articulation of mass culture as feminine was, according to Huyssen, part of a project to reaffirm the boundaries between high modernism and low mass culture and served to establish social hierarchies that maintained gender distinctions and perpetuated a patriarchy. As a result, where women in the nineteenth century increasingly accessed social spaces through the fields of work, leisure and technology and where women increasingly gained social power through women's suffrage, in many ways women still remained subordinated through their association with mass culture. And where popular culture was made available to the masses through technologies of production and communication, there was a correlating anxiety about the effects of this, anxieties that were sometimes expressed through discourses of feminization. Woman was seen as not quite belonging in these new social spaces, such as the theatre, the cinema and the amusement park, but was paradoxically invited into and representative of these same spaces.

Despite this, the development of nineteenth-century consumer culture was in many ways structured around and dependent upon the invitation to women to embrace and take pleasure in commercial spaces. Women were directly addressed as consumers within promotional literature for theatre, vaudeville, fairs, department stores and other popular performances and events. Women were allocated specific spaces and times in which they could consume and view at ease. Women were allocated times to attend medical lectures;[54] they were offered 'ladies-only' matinee shows, 'ladies-only' dining or cafes[55] and even 'ladies-only' ice skating.[56] Places such as the department store offered women relative sanctuary from the streets and functioned as an extension of traditional feminine spaces and of the domestic sphere.[57] These were, as such, spaces which reaffirmed traditional notions of female propriety and decorum, for example, in the department store's aesthetic display of feminine clothing and goods. At the same time, they encouraged activity and social engagement typically withheld from women. The department store was, for example, a site where women's relationship with technology was nurtured. Noah Arceneaux had pointed to the role that the department store played in making radio accessible to women. Noting how early twentieth-century radio was a masculine technology, Arceneaux suggests that department store radio broadcasts, which advertised the in-store goods, addressed female shoppers, thus acknowledging and encouraging women's radio listening.[58] A similar practice was undertaken in later years when television stations broadcast a schedule of consumer-friendly programmes in department stores in the United States. Anna McCarthy finds evidence of advertising trade discussions in the 1940s about how television advertising might be used in-store to appeal to female shoppers.[59]

Women were both encouraged into such spaces by proprietors but equally chastized publicly for feminine behaviour. Women were often ridiculed or criticized for their inability to control themselves in such spaces. Women's shopping habits were often contrasted with those of men, whereby women were represented and imagined as out of control, indulgent and hysterical. This was part of the growing concern about the ways in which late nineteenth-century social transformations were informed by and related to concurrent transformations in the practices and ideologies of gender. These concerns were played out through public responses to women's new roles in public life.[60] Women's consumption was pathologized in popular and medical publications.[61] Newspapers and magazines mocked the frivolity of women's shopping. One newspaper, for example, joked about women's ineptitude for shopping. 'In a shop that is selling off, the majority of ladies (who buy anything) will be found to have bought what they did not want, and

not to have bought what they did want. It is very amusing to watch
them torturing goods and to watch prices torturing them.'[62] The
disparagement of women's shopping behaviour pointed to anxieties
about women's independence from traditional feminine spheres and,
more generally, women's initiation into activities that centred on their
pleasures and interests.

Equally, there was a double standard in regards to women's
attendance at sites of consumption such as the theatre, vaudeville and
film. Throughout the mid-nineteenth century, owners of entertainment
venues began to open up typically male spaces to women. Highbrow
theatres and other venues had long since encouraged female audiences.
They often took the form of the matinee performance that offered
unescorted women a segregated space in order to help them feel secure
and to protect their reputation. The night-time shows, in contrast, were
either largely male or mixed, in the sense that women were escorted by
husbands or family members.[63] By the late nineteenth century, then, such
venues were opened up to a much wider clientele. However, gender and
class politics were still at play since women were expected to attend at
certain suitable times. For example, vaudeville theatre owners extended
an invitation to women, particularly working women, by branding
their venues and performances as respectable and decent. For instance,
in the *Omaha Daily Bee*, the Eden Musee venue advertised spaces that
were welcoming for women and children including the Lecture Room,
Curiosity Hall and Bijou Theatorium. 'Ladies and children are earnestly
invited to attend the AFTERNOON RECEPTIONS, thus avoiding
the great crowds at night.'[64] Women were, therefore, anticipated as an
audience, but equally segregated from the main showings in response
to their perceived vulnerability. Daytime lectures and shows served the
dual function, then, of welcoming women to historically male-defined
spaces but also of ensuring that unaccompanied women were separated
from the general, default population of men. The promotional material
and advertisements were positioned as responsive to perceived female
preferences. However, this also functioned to suggest that women's
unaccompanied evening and night-time attendance was dangerous

and improper. Matinee performances often prohibited drinking and smoking as well as obscene or inappropriate behaviour and language.[65] This was both to prevent women from offence and to maintain sex-segregation.

The women's matinee shows, like the women's shopping activities, became the subject of condescension and ridicule by men who saw women's engagement with them as improper and unrefined. Throughout the late nineteenth century and early twentieth centuries, competing discourses of the female audience emerged and some resistance to women's theatre and film attendance was evident. In relation to women's attendance at London's West End theatres during and after the First World War, for example, Viv Gardner reads the backlash to female theatre audiences as part of a wider social concern about the independence afforded to women who were offered more employment opportunities as a result of the war.[66] She also points to the moral panic that arose in reaction to the mixing of young single women with soldiers, some of whom used the theatre as a space to meet. Richard Butsch suggests this backlash occurred earlier – at the turn of the century – and as response to anxieties about the destabilization of gender and class boundaries.[67] While high art performances of the orchestra or the opera were held to be a suitable day out for ladies, shows that were considered low-class amusement were thought to bring the worst out in women. The 'matinee girl' (the female fan of theatre or film and who attended matinee shows enthusiastically) was representative of this destabilization of class and gender norms and was the subject of much criticism during the late nineteenth and early twentieth centuries.

The matinee girl represented for some a lowering of the tone and intellect of the audience. She was less interested in narrative, form or performance and more infatuated with the actors and stars of the stage. She was accused of 'nauseating idolatry'[68] or of 'doing stupid things'.[69] She was driven by emotions rather than rational thought. The pleasure she took in idolizing actors and performances was vulgar and unbecoming of women. Butsch argues that the matinee offered 'entertainment for women rather than moralizing education for them and their children, an escape and a refuge for women from paternal oversight and domestic duties'.[70] However, while this might have been the case, the critical judgement of women-only matinee shows resulted in a distinct framing of women's pleasures, desires and interests as sentimental, tasteless and superficial. Some marvelled at this new member of the audience and sought to find value in her interest in the theatre, while also acknowledging her failings. The *Washington Times*, for example, dedicated an entire page to stories and letters about the matinee girl in 1903.[71] It praised the matinee girl in some ways but also mocked her naivety. While the newspaper celebrated the diverse range of women in attendance, it also criticized the matinee girls' 'foolish idolatry' and for disrupting the actors and other audience members. It cited a letter from a 'leading man' who implored the matinee girls to be sensible and refrain from disturbing the performance. The newspaper described the matinee girls as 'spellbound' and noted how, upon leaving the theatre, the matinee girl's 'memory of it all lingers to conjure up mischief and nonsense in her frivolous brain'.[72] Another newspaper cited actor Robert Edeson's view of the matinee girl as distinctly low-class. Those higher-class, more refined women were less prone to sentiment, more interested in the play than fanaticism and were more sensible than the 'silly, sentimental matinee girl'.[73] Edeson reported confidently that the matinee girl would be increasingly relegated to the cheaper theatres once the cost of a ticket was increased. The matinee girl, identified as 'the half-educated young woman, assistant, typewriter, telephone girl, or what not', was held responsible for the influx of 'low culture' to entertainment.[74]

Women were scrutinized not only as the audience but also as the performers. Competing discourses emerged in relation to the spectacle of the female body, which became as much a product to be consumed as any of the contemporary goods on offer. In theatres, vaudeville, variety shows, burlesque, magazines, advertising and photography, the female form was represented as a desirable commodity and an aesthetic object. Andrew Erdman and others refer to the 'spectacularization' of the female body, which resulted from the development of modern corporate, and male, governance instituted through the rationalization of industries, including that of popular culture.[75] The mass media and entertainment industries did not necessarily initiate this spectacularization of the female body but were instrumental in systematically normalizing it in the social imagination. Carolyn Kitch, for example, points to the role that popular magazine illustrations played in generating and perpetuating stereotypes about womanhood and femininity around the turn of the century.[76] The female image in magazines like *Colliers* and *Life* functioned to sanction and legitimate certain types of femininity and womanhood, as well as class, national identity and race. Such magazines, along with other mass media and entertainment, normalized the spectacularization of the female body. This process of normalization was enacted through the codification of the female body in terms that quite often centred on sexualization. In commercial entertainment, the female body was produced as the object of consumption. In the marketing and promotion of consumer culture, the female body became the site through which desire for commodities was articulated.

Since the early decades of the nineteenth century, popular entertainment venues had experimented with risqué female displays of bodies and suggestive performances. Given the dominance of Victorian sexual morality, these tended to be subject to closure by the police and were targeted by moral reformers.[77] Yet the success of these shows, in which female dancers and singers would lose some of their heavy Victorian dress, encouraged venue owners to continue to provide such entertainment. Thus, the nineteenth century saw an increase in 'feminized spectacle' whereby the body of woman became a legitimate object of display in commercial entertainment.[78] While there was a decrease in indecent female display in seedier establishments such as concert saloons, this was met with a corresponding increase in displays of female sexuality in more upmarket and legitimate entertainment.[79] Moral reformers and the press initially resisted this sexualization of the female body, but by the latter half of the nineteenth century, such entertainment was widely accepted and attended. By this time, the female body had become a site of male scopic pleasure in commercial entertainment.[80] Thus, while entertainment offered women access to a field of labour ordinarily excluded from them, the conditions under which they performed were dictated by a repressive regime of sexualization.[81] This tension about female labour and sexual representation in entertainment culture can be seen, in part, as a tension between a more traditional and conservative model of womanhood that emphasized her purity and domesticity and a modern, more independent and confident woman who was defined by her overt presence in public and consumer life.

This 'New Woman' of the United States and Britain was educated, independent and more active in taking up many of the new forms of entertainment, sports and lifestyle opportunities offered in the late nineteenth and early twentieth centuries. She was, in equal measure, dismissed (in the press) and championed (in the Gibson Girl illustrations of *Ladies' Home Journal* or *Collier's Weekly* that presented the modern woman as assertive and independent) for her progressive ambitions. She represented a rejection of the mid-nineteenth-century discourse of True Womanhood, defined in terms of modesty, tradition and domesticity.[82] Yet there was little consensus about what modern womanhood should look like. Various forms of womanhood were continuously articulated and debated in public, in the press, in theatre, in film and by both men and woman. Figures and representations like the Gibson Girls or the chorus girl spurred debates about what the contemporary woman was and how she should participate in public life. For example, Linda Mizejewski argues that the chorus girl was a symbol of the liberated woman, of emergent forms of female sexuality and of the commodity culture that femininity would be identified with. Where the actual practice of being a chorus girl might have been governed by exploitation, cheap wages and a competitive labour market, the representation of her was marked as aspirational in the field of entertainment and as dangerous by reformers.[83] Ultimately, the mediatization of women and women's bodies was productive of a range of discourses that spoke of them, that shaped public opinion about womanhood and that sanctioned, but also managed and regulated, the participation of woman in the sphere of commercial entertainment.

During the late nineteenth and early twentieth centuries, then, gender practices and gender discourses were being negotiated in the popular press and in public entertainment venues. Women did come to be seen as important players in the new spheres of media and entertainment; however, this was conditional and subject to ongoing debate and critique. Women were afforded greater access to and participation in the new media and entertainment spaces and experiences; however, this was often conditional upon a set of expectations that were determined by producers, business owners, advertisers and the wider press. Early moving image, for example, came to 'impose masculine forms of vision'[84] through the development of a form of narrative and spectator address that prioritized the male gaze.[85] Scopic pleasure was generated through the female form as a consumable object of desire, thus reinforcing her association with commodity culture. Since it was men who controlled and commercialized media and entertainment, it was through them and their notions of female consumption that women were invited into the consumer economy. Women found a route to public life through the consumer sphere, rather than the sphere of invention and production, and had opportunities to articulate and form identities and lifestyles through the consumption and use of technologies of communication and entertainment. At the same time, the consumer sphere was regulated and governed by (largely male) business interests that perpetuated a hierarchy of sexual difference as much within the consumer sphere. By the turn of the century, women were far more visible and active within the consumer economy, as both workers and consumers.

This differentiation and segregation of men and women's production and consumption extended into the development of new forms of mass communication, including television. As with the previously mentioned media and entertainment, television technology was initiated within a distinctly masculine domain only to be later positioned as hetero-social when it became a mass medium and its commercial potential was realized. Once again, however, women's relation to television, as with other media, was governed by commercial and economic interests. In the next chapter, I account for the means by which women were positioned, not as producers, but as both consumers and objects of consumption. Women were imagined and addressed as domestic users of television, albeit ones who lacked the technical expertise of men. In addition, women were used as the 'face of television', a promotional role that again associated them with television consumption more specifically and consumer culture more generally.

Television's earliest years

In this chapter, I examine the early years of television as it was undergoing constitution as a service, a form of entertainment, a domestic and public medium. The chapter reviews the 'false start' of television from the early 1920s to the years of the Second World War where television was shifting between an experimental and a commercial medium, where it was proposed and, in some cases, offered as a public form of news and entertainment. I consider the ways in which women were imagined as part of television culture and as its audience, the ways in which they participated in its formation – as producers, announcers, purchasers and viewers. I begin by discussing how television demonstrations that took place in department stores, offices, laboratories and in public lectures were highly gendered affairs. During the early 1920s and into the 1930s, television inventors presented television to the public in settings such as Selfridges in London,[1] Proctor's Theatre in New York[2] and the New York World's Fair.[3] These demonstrations worked to situate women as the audience of television or as the objects to be consumed on the screens of television. Indeed, women came to have an important aesthetic function in regards to the promotion and marketing of television. Throughout the 1920s and 1930s, much attention was paid by broadcasters and the press to the appearance of women on screen. This materialized in the role of the 'television girl', often referenced in the trade press, and that would represent the visual qualities and value of television. In this sense, women functioned as spectacles of consumerism since their role was to create desire and demand for television.

As television increasingly took hold as an idea and a possibility, press and industry discourses worked to imagine women as part of the new television landscape and to shape women's participation in it. Within the realms of scientific practice and discovery, television was largely a masculine pursuit and governed by masculine authority. Ideas about television were framed by social conventions and practices of the day and women's role in television was governed by prevailing gender norms and expectations. Women's position in, and relation to, television was negotiated cautiously and not without tension. Women were predominantly imagined as having specific investments in television (either as working in television or as consumers of television). Concerns about women's technical knowledge (or lack of it) arose in technical magazines and popular newspapers.[4] Where women were to appear on television broadcasts, much was made of their appearance. And where women were imagined as viewers, their relationship to television was sometimes trivialized. Despite this, women's role in the development of television as an industry and practice was crucial to its success. Women both shaped and were shaped by the new medium in these formative years. Where the later institutionalization of television produced a specific idea of women as consumers and audiences, the experimental years of television reveal the ways in which this was formed and, in some cases, contested.

While the 1950s are widely considered the formative years of television, the preceding decades of the 1920s and 1930s were key to the eventual formation of the television industry as it materialized in the United States and Britain in the post-war years.[5] For William Boddy, the rise of the television industry did not occur seamlessly. Although television was a close relative of radio, much effort was made to distinguish radio from television. In the early scientific development of television, broadcasting was only one of many possible futures of the technology and questions of its utility were played out in public and private debates, in the press and by government and policy-makers. Broadcast television was by no means a given and its eventual realization in the post-war years was a result of much experimentation, innovation and failure. The proclamations about television's place in the media landscape stemmed from an institutional need to regulate or commercialize the medium as well as a public desire to integrate the new technology into contemporary life. Television was, indeed, imagined in ways that were at times realized and at other times unfulfilled. Where the word 'television' might have entered popular vernacular by the 1920s, its usage often reflected different ideas. Television could have followed a route towards personal communication, resembling more a videophone.[6] It could have become a consumer tool for shopping. It may have become part of the theatrical experience as was experimented with in the 1930s.[7] That it became a broadcast medium was not inevitable but the product of a complex range of debates, experiments and ideas.

A cursory history of the period suggests the extent to which television's future was very much up for grabs. During the 1920s and 1930s, television in the United States and Britain was undergoing experimentation. Inventors such as John Logie Baird, Lee de Forest, Charles F. Jenkins and Philo T. Farnsworth were experimenting with transmission and working to produce the definition of television technology. While there had been no formalization of technical standards as had happened with radio, inventors nonetheless promoted and demonstrated their systems to the public and the press. Given the technology's proximity to radio broadcasting, the United States and Britain saw some of the same companies in radio, such as BBC in Britain and NBC in the United States, expand into television, and experimental licences were granted to them and other companies and organizations by governments of both nations.[8] This allowed established broadcasting companies to not only test but also publicly showcase and promote their television technologies, resulting in many grand and ambitious announcements to the public. Many of these proclaimed achievements were premature, given the slow pace of television's development. This slow pace was partly due to the lack of technical standards: a necessity for the successful rollout of television in each nation. In Britain the BBC led the development of television, working firstly with John Logie Baird to develop a system and later (despite the investment of Baird) adopting EMI's electronic television as the technical standard.[9] The move to adopt a standard took place alongside the regulated scheduled transmission of television programmes from late 1936 to early 1947.

The speed at which technical standards were agreed and regulated transmission undertaken suggested that television was on the cusp of widespread delivery and broadcast. Between 1936 and 1939, the BBC continued to (somewhat sporadically) air a schedule of programmes, producing a range of content for the public, promoted in magazines such as the *Radio Times*. That television was accessible to very few viewers remained an issue, with the biggest barrier being, perhaps, geographical limitation of transmissions and reception, followed by the high cost of television receivers. This meant that while many were aware of television, and were indeed enthusiastic about it, not many regularly viewed it. It remained, during the late 1930s, a novelty.[10] In any case, by 1939 British television was suspended due to the war and the risk of broadcast interceptions by the German military. The development of television in the United States was equally governed by considerations of ownership, regulations and standards, with these issues being addressed in ways particular to the US governmental and broadcasting context. Like Britain, the US government intervened when television was realized as a technical possibility. While slower to establish a technical standard, the regulation of frequencies, issued through licences, resulted in a small competitive playing field, dominated by way of patents by RCA and its television subsidiary NBC.[11] In the years leading up to the eventual agreement on technical standards RCA and competitors like DuMont, Farnsworth and CBS demonstrated their experimental broadcasts to the public in an attempt to establish their place in the new television landscape and to ensure that they secured as much of the potential television market as possible.

It was within this environment of experimentation and presentation that emerging discourses of women and media materialized. Women had a number of roles in the television technical demonstrations. At times, they were the audience for the demonstrations. They were often recruited to act as performers on television, whether this involved sitting in front of the camera or engaging in song, dance or acting.

Although these demonstrations were primarily aimed at proving the viability of television, they were also beginning the work of defining the social and cultural form of television. The demonstrations ranged from simple presentations of images to large commercial showcases aimed at selling the idea of television as a product and experience. These demonstrations took place anywhere from offices and small studios to department stores, public conventions like Radiolympia in Britain and World's Fairs in the United States. Between the early demonstrations of technology in operation and the late 1930s marketing and promotion of television viewing as a leisure activity, ideas about television's utility and form developed. Initial demonstrations were concerned with the technical rather than the commercial potential of television (see Figure 3). These demonstrations were largely addressed through the masculine sphere of technoscience, whereby television transmission and reception were celebrated as achievements in and of themselves. Early newspaper

Figure 3 John Logie Baird (1888–1946), Scottish electrical engineer and pioneer of television, 1920s. Baird with one of his earliest experimental machines, with permission of Heritage-Images/TopFoto.

reports discussed the technology required for transmission and showed photographs of men operating complicated machinery.[12]

As television's potential as a public utility and commercial enterprise grew throughout the 1920s and 1930s, these demonstrations functioned to convince the public of its merits and pleasures and of the ways in which it would transform and enhance modern life. It is in this context that women were addressed as potential users and represented as among those who would feature on television screens.

The earliest public demonstrations of television – those that were aimed at a non-scientific audience – sought to shape the potential use of the medium. It is perhaps telling that one of the first public demonstrations of television took place in a department store, signalling the possible commercial role that television would play and entrenching it in a culture of consumption. The introduction to and demonstration of television in a department store also educated shoppers and the public more generally on new technologies and media.[13] By situating television technology in a familiar environment and context, women shoppers could see how it would fit into their everyday lives. In early 1925, Baird showcased his technology in Selfridges department store in London.[14] A rather crude introduction to television, given that what was displayed was a silhouette (what Baird termed 'shadowgraphs'), it nonetheless worked to imply that not only was television quickly developing but that it would be associated with the commercial sphere and would function as an extension of the lifestyle of those who could frequent department stores.[15] In the context of a department store, people were addressed as consumers and invited to consider television as one potential modern convenience available for purchase (with Baird's television receiver becoming available for purchase in the store in 1928).

In the United States, a more formal television demonstration by Jenkins in 1925 that was heavily promoted and publicized worked much harder to stress the public utility of the new technology. A group of representatives from government and federal agencies as well as reporters from leading US newspapers saw Jenkins demonstrate television images of a windmill turning.[16] Jenkins used the opportunity to champion the technology and shape ideas about how it might materialize.

> Folks in California and Maine, and all the way between, will be able to see the inauguration ceremonies of their President in Washington, the Army and Navy football game in Philadelphia ... The new machine will come to the fireside ... with photoplays, the opera and a direct vision of world activities.[17]

Jenkins framed his 'radiovision' as a domestic medium and technology, positioning it as one that would bridge the divide between the public and private realms. For Jenkins, television was a democratic medium that would

> come to the fireside as a fascinating teacher and entertainer, without language, literacy or age limitation; a visitor to the old homestead with photoplays, the opera and a direct vision on world activities, without the hindrance of muddy roads or snow blockades, making farm life still the more attractive to the clever country-bred boys and girls.[18]

Noting the way that radio had changed the 'social order', Jenkins pointed to how millions could listen to great men *and* women. For Jenkins, television could re-shape social hierarchies. For him, the potential of the medium was in its accessibility to all ages, social classes and genders. The idealism inherent in Jenkins' imagined future for television was accompanied by a firm certainty of its impending arrival.[19]

This social shaping of early US television occurred throughout the 1920s and into the 1940s where many different potential uses were proposed. Noah Arceneaux suggests that department store television systems – which existed from the 1920s to the late 1940s – evidenced alternative possibilities for the new medium.[20] Department store television was far more commercial than US television went on to be; however, it also addressed women more directly through the television programmes that promoted goods. The New York World's Fair television demonstrations by RCA included experimental programmes for and about housewives, such as domestic comedies.[21] RCA also constructed model living rooms that suggested how television might be accommodated in the domestic space. As a key market, female audiences and consumers were at the forefront of television's address and were normalized as potential users of the medium.

Women may have been imaged as equal users of television; however, their involvement in the practical demonstrations of television was governed by gender norms. Women would have important supporting roles as hosts, performers or switchboard operators. This was the case, for example, in an American Telephone & Telegraph (AT&T) demonstration of long-range television transmission on 7 April 1927 in New York and New Jersey. Despite being one of the first demonstrations carried out by a large telecommunications company,[22] and one intended to promote the commercial potential of the medium,[23] it was accompanied by cautious optimism at best. Drawing from the form of radio broadcasting, and in order to garner widespread public attention, Secretary of Commerce Herbert Hoover was invited to be telecast. Guests were invited to participate in the demonstration and to enjoy a series of entertainment pieces, including a vaudeville act, a radio programme of 'a short humorous dialect talk by a lady'.[24] This was, indeed, one of the first public demonstrations to incorporate experimentation with the form of television as well as the technology. The programmes (or 'displays' as they were called) that made up the demonstrations reveal much about how television would later materialize as news and entertainment. The *New York Times'* report on the demonstration suggested the importance of the television performer's looks and beauty. The report, for example, began with a description of the technical delivery of Hoover's performance and then moved on to discuss the second set of performances which were formed of variety-like pieces. An engineer who first appeared, it was claimed, 'has a good television face'. The *New York Times* writer went on to address the vaudeville act in which an 'A. Dolan' performed a series of ethnic stereotypes for amusement, playing upon the visual and verbal otherness of Irishness and blackness by covering his face with dark makeup and a false moustache.[25] This form of entertainment was considered ground-breaking, with the comment that 'In its possibilities it may be compared to the Fred Ott sneeze of more than thirty years ago … For the commercial future of television, if it has one, is thought to be largely in public entertainment.' The female comedian was given less attention with her performance, receiving only passing comment. This echoes what Leon Rappoport describes as the tendency to defer to those perceived as in authority even in the case of comedy.[26] During the 1920s and 1930s, those female comedians working the comedy circuit in the United States often played a secondary and supporting role to male comedians.[27] This dynamic is perhaps reflected in the article.

More was made of another woman, the switchboard operator named Edna Mae Horner, who connected the calls between New York and Washington. 'This one was a good-looking girl with fluffy hair, and as cool and efficient as if she had been at the television-telephone switchboard all her life.'[28] This comment underscores the complex status women would have in the new mass medium. Women were acknowledged contributors, noted for their professionalism and skill, but ultimately bound to conventional gender discourses that valued their appearance. The switchboard operator in this AT&T demonstration was noted for her contribution to the successful coordination of the test. The role of the male engineers, politicians and performers was assumed as professional and this was no surprise given the association between masculinity and technical and professional capability. Skilled men were standard; skilled women were exceptional.

The reference to the ability of the switchboard operator suggested the participation of women in the professional labour market but also suggested the ways in which these roles were scrutinized and discursively produced. By the 1920s, certain communications roles had become commonplace for women and the female switchboard operator formed part of the new media labour economy. This job enabled women to participate in skilled labour but also depended upon women enacting the social duty of care and maintaining a polite manner during telephone exchanges.[29] This switchboard operator, discussed in the *New York Times* article, was reflective of a role common among young educated women who not only undertook skilled labour but also necessitated a distinct presentability, much like the telephone switchboard operators employed by Bell and other companies.[30] The *New York Times* author's emphasis on the pleasing looks of the operator suggested an attempt to negotiate an appropriate understanding of the televised woman.[31] It at once celebrated her televisuality but also described her televisual qualities according to conventional gender-prescribed notions of beauty. Where the engineer was noted as having a face for television, the operator was noted as having a beauty that transcended television. Hilmes notes a similar tendency towards objectification of the female radio operator, whereby the threatening presence of women in a supposedly masculine field was somewhat negated by reference to her appearance and her physical attributes.[32]

The AT&T demonstration pointed to the type of programmes and schedules that would define television and, during the late 1920s, there was far more evidence of experimentation with a range of programmes and genres as well as with more regular broadcasts. General Electric's broadcasting station, WGY, began a regular schedule of transmission in the United States in 1928 and other stations started to follow suit.[33] In Britain, the BBC, having been reluctantly tasked with bringing television to the nation, transmitted its inaugural service in September 1929 even though, as Burns notes, receivers were not yet widely available to the public for purchase.[34] These transmissions, like in the United States, drew upon the formats and genres of radio. By the time the BBC transmitted its first televised play in 1930, television receivers were on sale but sales were low and few witnessed the transmission. Despite this, by the early 1930s programme schedules in both Britain and the United States revealed a variety of content that resembled that of other popular news and entertainment forms. The BBC, for example, televised the play *The Man with the Flower in His Mouth* in 1930,[35] followed later by a transmission of the Derby horseracing event which was the first outdoor broadcast for BBC television. The BBC experimented with a number of programme formats. In the United States, there was a similar move to programme schedules with major radio companies such as CBS and

RCA (through NBC) offering short regular transmissions. Among CBS's programmes were boxing matches, football games, news and politics, and variety and entertainment. Mike Conway suggests that since CBS was not interested in receivers, its attention was on the development of television content and the television experience.[36] Indeed by the time it suspended its experimental television efforts in 1933 (due to the Depression), CBS had a sophisticated schedule of popular culture programming that exploited the best of existing entertainment media. Still clearly experimental, these tests served to structure an institutional culture of television production and form. In its progression from a laboratory experiment to an actual service, questions about what it might be, and who would participate in and form this new television ecology, materialized. As more television programmes were made and as more press attention was paid to the medium and its content, there were more roles for women.

What these roles might be was determined throughout the 1920s during which time clear patterns emerged in regards to the nature of women's participation in television and the function women would serve, particularly as ambassadors of sorts for television. During the 1920s, clear patterns of inclusion and exclusion would emerge. Although the role of women as the television audience would eventually come to dominate discussions of women's access to and relationship with television, earlier concerns pertained to women's involvement in television production and programming. During the 1920s, both radio and television were mediums under development and it is no surprise that some of the opportunities for, and concerns about, women in television stemmed from the experience of women in radio.

In a pattern that was to anticipate the television industry, radio began as a space that offered potential roles and opportunities to women but these became limited as time went on. According to Susan Douglas, women's voices had a place on radio in its earliest years.[37] Women were also able to secure work in radio production and across a wide range of roles in the broadcasting industry. For Kate Murphy, BBC radio at least provided some opportunities for women to work in radio production.[38]

The BBC had a more enlightened approach to women's employment than was typical in other organizations and companies at the time in Britain (despite the passing of the Sex Disqualification Act of 1919, which was to offer women access to employment). As a new organization with no legacy of gendered working patterns, women could sometimes progress through the ranks to senior positions. Hilda Matheson, for example, was headhunted in 1927 from a role in politics and became Director of Talks at the BBC. Mary Somerville joined the BBC and became Director of Schools Broadcasts in 1929.[39] Isa Benzie joined the organization in 1927 and became Director of the Foreign Department by 1936.[40] Mary Adams joined the BBC in 1930 producing radio programmes on the topic of sciences, before moving to television talks in 1936.[41] In the United States, women were also present in the earliest years of broadcast radio. Eunice Randall began her career producing technical drawings for radio receivers in 1918 and her interest in radio led her to become one of the earliest radio announcers.[42] Bertha Brainard found work as a volunteer with station WJZ in 1922, moving to NBC in 1926 and becoming one of the first female executives at the organization.[43] Judith Waller began her career with station WMAQ in 1922 promoting public service programming and going on to a career at NBC as Public Service Director in the company's Central Division.[44] Gertrude Berg began her career in radio translating advertisements into Yiddish but soon shifted to creating her own programmes, with the iconic programme *The Goldbergs* (1929–46; 1949–56) broadcasting first in 1929 on CBS radio and moving to television in 1949.[45] Mary Margaret McBride began her radio career later, in 1934, but quickly rose to become one of the most success broadcasters, hosting magazine-style programmes for CBS and later NBC.[46] From the formative years of radio, then, there were some pioneering and trailblazing women broadcasting and women were audible to the radio public.

Hilmes notes, however, that initial opportunities for women as producers and announcers were eventually restricted to gender categorized genres and gendered timeslots (with daytime television becoming synonymous with the housewife).[47] Those aforementioned pioneers formed the exception to a general rule of gendered radio. Hilmes identifies the concerns about women's voices on radio, whereby the female announcer was not considered authoritative or commanding enough and, thus, excluded from announcing roles. She cites the debates among station owners and listeners about the different qualities of male and female announcers and the tendency to question the legitimacy of the female voice. In the context of television announcers and talent, similar concerns were articulated in relation to women's appearance. Unlike the experience of radio, however, television was generally accepted as an appropriate site for women, particularly given the precedents set by female stars of the theatre and cinema.

It seemed inevitable that women would appear on television programmes as hosts, announcers and talent even though the types of duties and responsibilities they would undertake would be determined by social conventions of gender. In the United States, the press tended to centre on the exciting new role of the 'television girl', a role that saw women act as ambassadors for and visual representatives of the new medium. In Britain, the BBC's organizational structure and progressive ethos allowed some opportunities for women, with the public service status of the BBC resulting in consciousness about the participation of women and the role they would play in radio and television.[48] In both cases, wider public discourses about women in television centred on the nature of women's work, the roles they would undertake, the qualities a woman would need to be employed in television or appear on television and the value television might bring to women's lives.

One of the qualities that a woman could bring to television was her stardom. The female star, in this context, convinced the viewer of the legitimacy of television as an entertainment form. As Susan Murray suggests, television 'used its stars to define itself'.[49] The female star – already established in other media – worked to persuade the public that television could be a popular arts and entertainment form. While few would have witnessed the female star on television in the 1920s and early 1930s since sales of television sets were so low, there were plenty of opportunities to read about her television performance and to see photographs of the star in the television studio. Theatre actresses such as Dorothy Knapp and Izetta Jewel were recruited in the late 1920s and 1930s either for television promotional activities, to star in experimental television programmes or to assist in demonstrations. Competitions were held to find the new 'television girls' at pageants, in the press and at radio and television stations. At the 1939 New York World's Fair, a contest was held for the next 'television girl' and contestants were televised and broadcast on the NBC experimental station (see Figure 4).

The press featured stories on what the women's roles were in television and published photographs of the women posed in the television studio, near television equipment or, in many cases, posed for publicity stills with no visual reference to television.

Figure 4 Woman being filmed in costume at New York World's Fair 1939, with permission of *New York Daily News* Archive/Getty Images.

For example, a 1931 issue of *Radio Digest* titled 'Television Is Here' introduced the female stars who would be the new faces of television.[50] Each star had a full or half-page publicity still and was presented as a potential cross-over talent, from a medium such as radio into television. The photographs of the woman foregrounded their physical beauty, with the women posed in such a way as to draw attention to the neck, legs or body. A short photo caption explained the qualifications of each woman that made her suitable for television stardom. Vaudeville actress Frances Williams was said to have a 'splendid record' as an actress as well as 'televisibility'.[51] Bernadine Flynn and Georgia Backus were both radio stars who were figured to be capable of making the transition to television. Backus, while noted as a director, was praised for her physical qualities: 'Those Television eyes!'[52] Dorothy Knapp and Natalie Towers were referred to as Miss America and Miss Television respectively, thus fixing their identities to pageants and displays of physical beauty and femininity.[53] *Radio Digest*, therefore, worked to define the meaning of television through female stardom and to define the meaning of women in relation to television. Television, like the female stars, was something desirable. It was enticing and alluring. The role of women on television was to represent these visual qualities and the female star was the vehicle through which to achieve this. Significantly, most of the work done in cultivating this relationship between women and television occurred before the medium was publicly available.

The female star, therefore, played an important role in securing television's status as a popular media in the early promotion of television. Television demonstrations capitalized on women's appearance and presentation in order to demonstrate the aesthetic qualities of the television image. Murray suggests that early experimental television-makers believed that performers such as actors, comedians and singers would emphasize the 'visual nature of the medium'.[54] The General Electric demonstrations, for example, included the televising of plays and other musical and comedy performances and used female stars of radio and vaudeville. Alongside these performances, General Electric released press statements and reports pointing to how television might be used within the home. An April 1928 General Electric demonstration televised the play *The Queen's Messenger*, a twenty-minute feature performed by Izetta Jewel and Maurice Randall. Jewel's currency as a stage star as well as a politician was capitalized upon with a press release announcing her as 'the first leading lady of a play presented for television'.[55] An *Associated Press* image had her posed as though acting in front of the camera.[56] By presenting the still-experimental technology through the prism of popular entertainment, the demonstration helped to promote its utility and its pleasures to the public. Television was positioned as another entertainment media, one which offered familiar formal and generic qualities like that of radio and film as well as the novelty of sound with vision. Jewel's presence functioned to give the demonstration legitimacy and to bridge the gap between television as an experimental apparatus and television as an advancing medium alongside radio, cinema and theatre. It also formed an association between women-as-objects-of display and television, whereby women's performance and spectacle would operate as the enticement for potential television viewers.

When women appeared on early television demonstrations, for example, reference was often made to their appearance and dress rather than their competencies and skills as announcers, presenters or talent. Women's 'softness' and general 'feminine' attributes were considered both a gift and a hindrance. A 1931 newspaper report on the types of 'girls' suitable for television offered a confusing portrait of the ideal female candidate. She could be short and shapely but not baby-faced. 'There must be none of the baby doll femininity [*sic*] with its china-like molded features and startlingly blonde hair'.[57] Instead 'the ideal television girl' would 'have red hair, flashing eyes of any color, flawless teeth and bold facial characteristics'.[58] This medium-specific beauty required for television was reported on widely in the press, such that a sarcastic question from journalist Allene Sumner asked 'what would the commercial world do without pretty girls to demonstrate each and every necessary or inane invention?'[59] Like soft faces, soft voices were not considered appropriate for television. According to

1928 reports on a Baird television demonstration, the televising of a woman had to be dispensed with since her 'feminine tones' were not transmitted effectively.[60] Rather than considering this a technical problem or limitation, this was instead perceived to be an issue with women more generally. At another demonstration in 1929, it was noted that a woman's face 'came through indistinctly' which was suggested to prove that men's faces suited the medium better.[61] Elsewhere it was reported that the ideal 'television girl' required 'a broad face and big eyes'.[62] A woman's hair colour was also important to note, with Charles E. Butterfield of the *Associated Press* quoted as claiming that 'red-headed girls are the best for television, with the brown heads a close second. Blondes are not so good, because there is not sufficient contrast'.[63]

The ideal television woman seems to have been found by NBC in 1939. This was 'Miss Patience' – a mannequin dressed in a black satin swimsuit that acted as a stand-in for test shoots. Among her qualities was the fact that she never protested working long hours or complained about the heat lamps. The ideal television girl suffered in silence and smiled.[64] The concern about appearances did not extend to men in the same way, suggesting that the function of women on television was largely aesthetic and ornamental. This pointed to some of the expectations of women in and on the new medium and reflected the culture of beauty already established in other media such as magazines or cinema. In other words, it was anticipated that women would be central to the address of the television image but that this address would be governed by the discursive regimes established in other media. These questions of the suitability of women to television were addressed by emphasizing the aesthetic pleasures afforded by her presence – at once alleviating the threat of female authority and agency that the woman on the television screen might pose and defining the female television personality as an object of beauty.

The discourse of the 'television girl' evidenced the production of a safe and 'appropriate' role for women in television. Throughout the late 1920s to the mid-1930s, television services, stations, broadcasting networks and the trade and popular press used the television girl as a means of promoting and advertising the new medium and of suggesting the normative gender relations that would structure television programming. The television girl was an advertisement of sorts – a physical embodiment of the pleasant, friendly and inoffensive form that programmes would take. The television girl's function was to encourage the public to invest in television. She was contracted by a broadcaster to both feature on television screens and promote television in the press. An issue of *What's on the Air* from June 1931 referred to CBS's and NBC's appointment of television girls Natalie Towers and Dorothy Knapp and framed the work of the women in this way. 'The two young women register perfection, "televisionally speaking." They are the television girls of NBC and CBS, respectively, and will be a major incentive for folks to equip with short wave television sets.'[65] Towers was featured on the cover of magazines such as *Mid-Weekly Pictorial* where she was named as 'Miss Television'.[66] Dorothy Knapp already had a career in entertainment as a Ziegfeld Girl and had been called the 'most beautiful girl in the world'.[67] Thus, the television girls advertised the pleasures of television and those pleasures were rooted in the woman as spectacle.

Not only would women be used to promote television culture but they would also be used in television programming to promote other consumer items. In the 1932 book *The Outlook for Television*, Orrin E. Dunlap recognized the opportunities for television advertising and cited a case where CBS used Towers to showcase a number of rare pieces of jewellery from Cartier.[68] Dunlap considered this a successful demonstration of the possibilities of television as a vehicle for advertising whereby 'beautiful girls ... will be called upon to play leading roles in the television broadcast that advertise anything from coffee ... to the latest hats, shoes, dresses, pyjamas, bathing suits, cigarettes, candy and soup'.[69] In press references to the television girl, her physical characteristics took precedence over skills or professional responsibilities. Magazines and newspapers gave detailed attention to the expected physical attributes required of the television girl, including optimal hair colour, facial characteristics, height and tone.

Some saw opportunities for women who were 'plump and over forty' since the poor quality of the transmission meant that women normally excluded from film might find their physical characteristics obscured enough to warrant their presence on the television screen.[70] Regardless of whether television was more 'forgiving' or not, women's presence on television was discussed more in terms of appearance than of talent. And while the suggestion that television might be low quality enough to allow for a range of physical 'types', in reality women on television, particularly the television girls, were typically young and slim and fit standard models of beauty. Drawing from the conventions of stardom and celebrity already established through other media, the television girl was, to paraphrase Richard Dyer, part of the way that television was sold to the public.[71]

In order to develop a television industry and culture separate from radio and cinema, it was important for broadcasters to cultivate television stars and personalities. The television girl served this purpose, with photo opportunities and press releases announcing both her status as the new television girl of a particular station or network and the entry of that particular network into the television industry. This was quite important given the slow development of television and its peripheral status in relation to radio and cinema. Television was at risk of being perceived as an extension of radio and cinema. Equally, the delayed deployment of television as a mainstream technology risked causing frustration or disappointment in the public. The television girl was to sustain interest in television during its slow development and operated as a signifier of the uniqueness of television. Television girl Hildegarde, for example, became an iconic representative of the emerging television star system. Already a popular cabaret singer in the United States and Britain (and who later went on to become a popular radio and television personality), Hildegarde was recruited to promote television to the public. From the mid-1930s to the 1940s, the press interest in her sustained an industry that was slow to materialize. Her appearances on television were often reported on despite the fact that there was as of yet no significant audience for it. Her persona was cultivated by referencing those televisual qualities that separated her from the world of radio and cinema. In *Radio Mirror* she was featured as one of the television stars 'soon to be famous beyond her wildest dreams'.[72] The feature established television first as a star vehicle:

> Perhaps you're wondering why television has not selected its stars from the famous ones already established in sound broadcasting. Instead it has hand-picked its first regular performers from the ranks of comparative unknown and sustaining artists and elevated them to the enviable status of being the original shining celebrities in a great new field – and these are undoubtedly the first people you'll see when you get a television set of your own.[73]

Television here was suggested to have the power to create stars. The manufacture of stardom suggested that the television industry was already in operation and working to fulfil the desires of both young starlets and audiences. The star system that television broadcasters sought to generate through the television girl was intended to guarantee its legitimacy as an industry.

This mythmaking continued throughout the *Radio Mirror* feature and used Hildegarde's role as television girl as a way of both distinguishing television from other media and demystifying its operations. The star acted as the vehicle through which the *Radio Mirror* reader would come to understand what to expect from television. This was important since, as the article seemed to assume, few had actually experienced any television. The article listed the 'special reasons' for Hildegarde's suitability as a television star. She had experience of it (having worked in French television). Not only could she sing but she could perform in a manner appropriate for television. Her 'gesticulations and facial expressions' fulfilled the visual expectations of television, making her more suited to television than radio. The article pointed to the limitations of radio where the projected voice of the performer failed to fully express the performance holistically. The article suggested the misleading nature of radio where romanticized stars might actually turn out to be less pleasing on the eye. Hildegarde was, therefore, presented as a more authentic star: beautiful in both voice and body. The article recounted Hildegarde's education in television, underscoring the learning process undertaken by her, whereby she was required to manage her performance for the camera and adjust her costuming to meet the demands of television studio production. The narrative of Hildegarde's involvement in television served to outline to the reader how television was produced and how the industry was evolving. Representing an industry not yet realized, Hildegarde functioned as an interim commodity for a television market not yet providing the tools and content necessary for television consumption. Magazines and periodicals provided a proxy form of television service in which television culture could both materialize and be consumed and the television girl acted both as a gateway to the new television product and as a product of television.

Ultimately, then, television slowly emerged as a gendered medium. While experiments with television aimed to address practical and logistical issues of transmission and reception, they also worked to determine the place of women both on and beyond the screen. In certain spheres and to certain audiences, television was presented as an open, hetero-social medium that could accommodate female engineers, technical workers and employees.[74] However, the reality was that women functioned mainly as photogenic advertisements for television. Their most visible role within the emerging television industry was not as producers but as representations of television. Although this was a highly public role, it was largely passive. Women were positioned as one of the pleasures of television. The television girls did gain attention, but mainly in regards to how presentable or 'televisual' they were. Nonetheless, as experimental television production became more regular, some women did manage to find a route into it. In both Britain and the United States, women found some opportunities available in the emerging industry. The following chapter identifies some of the work and experience of women employed in British television at BBC television during the period of 1936 to 1939, when regular experimental broadcasts were carried out.

Women in early British television

This chapter considers the roles played by women in the establishment of BBC television during its experimental years. In comparison to other large organizations and companies in Britain, the BBC claimed to encourage a more hetero-social environment across its broadcasting services and operations. Indeed, the first announcers of BBC television, Jasmine Bligh and Elizabeth Cowell, were women. They introduced BBC television to a small but enthusiastic audience and it wasn't without remark that they were women. In addition, women were employed in television production and helped shape the form that television programmes would take once television experiments ceased and the television service began in the post-war period. Producer Mary Adams, for example, experimented with a range of factual and educational formats that had much influence on post-war news and factual television. However, despite the egalitarianism of the BBC, many roles and responsibilities were gendered. This was evident in the way that Cowell and Bligh, despite their responsibilities as announcers, were also expected to serve the same function as the television girl discussed in the previous chapter. Throughout this chapter, I examine the roles that these women played in the development of the experimental television service. I consider the way the BBC and the British press negotiated women's presence in television and how this impacted their roles. I argue that, while these women were understood largely in relation to their gender, they nonetheless evidence a period in which women were key figures in and producers of British television. It is particularly important to recognize the participation of women during these years since it problematizes a general historical narrative that imagines that women's presence in television has been on a general upward trend.

This scholarship, then, contributes to more recent feminist revisions of British television history that trace the roles and challenges of women in BBC television over the past century. In feminist historiographies of British broadcasting, Kate Murphy's *Behind the Wireless: A History of Early Women at the BBC*[1] accounts for the multitude of women working in BBC radio and television in their formative years. Although Murphy focuses more on women in radio during the years 1936 to 1939, her work helps explain the organization, general operations of the BBC and the overall work of women there. In radio, as mentioned in the previous chapter, women secured positions in producing and announcing during the 1920s. Women were also employed as experts or public intellectuals as was the case with the initial employment of Mary Adams (who is discussed in this chapter) or S. Margery Fry, who took part in discussions on BBC talks and current affairs.[2] In addition, women were addressed as an audience through programmes such as *Women's Hour* (1923),[3] *Household Talks* (1927)[4] or *The Week in Parliament* (1929),[5] the latter of which was proposed by Hilda Matheson specifically for newly franchised women (in 1928 women over twenty-one had gained the vote in Britain).[6] Despite these gains, Murphy identifies a number of patterns and contradictions in the BBC's employment of women and of their opportunities and careers in the organization. She claims that, while the BBC was in many ways a progressive organization in its employment of women, it nonetheless differentiated the types of contracts and roles women could have. Waged and salaried staff were treated differently, with waged women subject to gender stereotyping in terms of the roles they could take.[7] However, Murphy suggests that in the post-war years, the BBC's attitude towards women changed with some departments and areas becoming 'increasingly hostile to salaried women', including 'News, Outside Broadcasts and Light Entertainment'.[8]

For Mary Irwin, the BBC's turn away from women resulted in their near erasure from the BBC's historical narrative.[9] Irwin cites television producer Doreen Stephens' lack of presence in the history of television less as a testament to the quality of her work and more a problem of archiving, whereby women's programmes were less likely to be preserved. This 'disappearing' of women's programmes and, as Irwin points out, the Women's Department at the BBC followed a path whereby women's programmes were increasingly sidelined both in the television schedule and across the organization. For Irwin, this resulted not only in the disappearance of women's programmes at that time, but a lack of focus on women's contributions to BBC television in later years.

Nonetheless, some women achieved significant status and acclaim in the post-war years. Women such as Grace Wyndham Goldie and Freda Lingstrom became iconic pioneers in the fields of current affairs and children's programmes, respectively.[10] However, this was in an environment that no longer valued the work of women as it did the work of men. Equally obscured from early television history was the experience of other female production and technical workers. Emma Sandon has mapped this history and accounted for the experience of female technical workers at BBC Television during the 1940s and 1950s.[11] Sandon paints a picture of a work culture somewhat hostile to female technical assistants, particularly after the war. Kathryn Terkanian, like Sandon, finds women's route into technical and production roles more difficult after the war, although she does note that some of the women recruited in the war years found other roles within the organization.[12]

Across the BBC, then, television production and technical work became gendered at pivotal points in the organization's early history even while it was more open to women across the workforce. As the BBC professionalized, roles, contracts and wages or salaries determined the extent to which a job was gendered. However, windows of opportunity emerged for women at various points, most notably in the war years. The lack of suitable male employees enabled women to take up 'masculine' roles in production and technical work. Many women continued in these or adjacent roles in the years following the war, when the BBC could recommence with the recruitment of men. Therefore, the women who succeeded in gaining employment in roles more typically understood (or explicitly identified) as male formed a large group of exceptions to a generally held rule that viewed such roles as unsuitable for women. This has resulted in a popular history of the organization that foregrounds men's work and disappears women's work. The literature cited attempts to revise this history and account for the work of women across a range of production, administrative, technical and managerial roles. In retelling a history of television inclusive of women, this helps frame our understanding of the current television industry and women's place in it. Yet largely absent from these otherwise rich histories is the role of women during the earliest years of BBC experimental television. While the BBC television department was considerably smaller in the pre-war years, it nonetheless had a fairly high representation of women in key roles. For the remainder of this chapter, I examine the roles undertaken by women in this department prior to television's cessation with the commencement of war in 1939.

During the late 1920s and early 1930s, the BBC was closely monitoring developments in television. John Logie Baird had first begun carrying out a number of individual tests and experiments with transmission but the BBC was initially far from enthusiastic about his mechanical television.[13] With some reservation, the BBC allowed Baird use of its station for the purposes of experimental transmission beginning in 1929.[14] During the early 1930s, the BBC remained discreet and cautious in making any announcements about its intention with television, with Director General John Reith requiring the BBC publication *Radio Times* to avoid reference to television without first getting clearance from senior BBC figures.[15] Throughout the early 1930s, the BBC broadcast a fairly regular schedule of experimental programmes such that '[b]y 1932, the BBC probably had more programme-making experience in TV than any other broadcaster in the world'.[16] Therefore, when a committee (the Selsdon Inquiry) was set up in 1934 to decide how television would be delivered and by whom, the BBC was selected to provide television to the public.[17] The BBC then began to plan for the launch of a regular experimental television service for commencement on 2 November 1936. This necessitated an internal and external recruitment drive that would result in the employment of a number of women in the newly formed television department.

One of the most widely promoted positions was that of the television announcer. The decision to appoint female as well as male announcers was, according to television critic Kenneth Baily, a conscious decision on the part of Director of Television Gerald Cock.[18] Baily suggests that Cock very intentionally employed women in order to exploit their 'warmth' and 'glamour'.[19] Since television was a visual medium, it needed something, or someone, to entice the viewer to the screen. The female announcers would serve this function. The BBC heavily promoted the new job of female announcer in the British press. Word of the new role was even reported to have travelled as far afield as Australia where the *Sydney Morning Herald* led with the caption 'Television Announcer: Search for a "Superwoman"'.[20] That the BBC would hire a woman as the first announcer demonstrated its progressive ethos, but the reference to her being a 'superwoman' indicated that much more stringent criteria be attached to her role than that of a man. The qualities listed did not seem to suggest any capabilities beyond what an ordinary woman might have and were largely related to presentation.

The BBC was evidently somewhat concerned about having a female announcer, not because men might find it objectionable, but because women would do so.[21] The *Sydney Morning Herald* stated:

> The BBC is conducting a search for a Superwoman. She must have a super personality, charm, tact, a mezzo voice and a good memory. She must be as acceptable to women as to men. She must photograph well. She must not have red hair and must not be married.[22]

Back in Britain, newspapers reported that candidates were to 'have good looks, personality, and a perfectly pitched voice'.[23] The tone of the press announcements was both paternalistic and patronizing, acting as a form of disciplining of the prospective announcer. The criteria against which the female candidates were judged differed from that of their male counterparts. It was reported that the

> male announcer must have a clear, cheerful voice ... average height, well-proportioned and unprominant [*sic*] features; preferably dark eyes and no red hair. Good education and good memory, previous

stage experience and ability to pronounce foreign words and names an advantage.[24]

Where the female announcer position was largely dependent upon extraordinary characteristics, requiring 'superwoman' abilities, the male announcer was required to have a clearly defined set of skills. The female announcer had to transcend the ordinary and be distinctive and outstanding where the male announcer was to identify with the 'everyman'.

This difference pointed to the reproduction of normative gender binaries that positioned men as productive and women as products. The male announcer was to act as a point of identification for the audience – their 'ordinariness' and 'good education' constructing them as the discursive voice of television. The female announcer would be largely ornamental, bringing aesthetic pleasure to the viewer. This was evident in Jasmine Bligh's claim that the BBC was more concerned with her body than with her skill:

> They [BBC personnel] wanted to know what our legs were like. Now we never showed our legs on television so, instead of saying 'Miss Bligh would you raise your skirt slightly', they made us climb up three steps and get onto a table where, no doubt, they discovered what they wanted to discover.[25]

This polarization of characteristics was further stressed in a lengthy news article in the *Yorkshire Post and Leeds Intelligencer* responding to the announcer positions.[26] Concern was raised over the implications of the visual representation of announcers. In a moralistic and condescending tone, the article outlined the need for the announcers to have 'an appropriately dignified and agreeable appearance'.[27] Contemplating the difference between a television announcer and a film actor, the article reiterated the need for the male announcer to express 'ordinariness'. Particular attention was paid to the indignity suffered by men subjected to female desire and fandom, whereby the male film star received 'somewhat embarrassing attention'.[28] This suggested anxiety about the unsettling of gender boundaries instigated by the mass media and the discomfort experienced at the display of female desire. The attention was inferred as unwanted and the objectification of men by women as abnormal. Where female pleasure in the male announcer was regarded as silly and unsolicited, pleasure in the female announcer was considered potentially dangerous and morally corrupting. Pointing to the trend in such practices in audience fandom and pleasure, the article asked 'must we not fear that the employment of female announcers in Alexandra Palace may strengthen immensely this disturbing influence, even to the point of disrupting family life?'[29] Women's engagement in television as either announcers or viewers was considered a social crisis likely to produce disorder and tension with the domestic space. The female announcer disturbed gender boundaries firstly by holding a position of influence and power over men and secondly by collapsing the distinction between the public and private realms. The female announcer not only jeopardized family life by potentially distracting the patriarch of the home but her position within the working sphere brought her outside of her prescribed role within the home.

Despite these voices of dissent in the press, the BBC in 1936 appointed two women as its first television announcers. They were Jasmine Bligh and Elizabeth Cowell. This recruitment of women might be accounted for by the promotion of the BBC as an institution committed to equality and diversity. In the months during which the recruitment process was underway the British press widely reported on the jobs for women announcers. The *Nottingham Evening Post* in July 1935 led with the headline 'B.B.C. to have women announcers?'[30] Although the headline suggested a certain amount of hesitation, the report itself referred to the advocacy of the BBC governor Mrs. Mary A. Hamilton for women's opportunities at the BBC. While some noted the 'severe specifications'[31] assigned to the female announcer job role, the position was nonetheless seen to afford good opportunities to women. Murphy points to the BBC's reputation for not only jobs but also parity in pay and promotion opportunities for women:

From the start, the BBC operated a non-gendered grading system for salaried staff who, in principle, were offered equal pay and equal promotional opportunities: practices which would have been music to the ears of women teachers and civil servants campaigning for parity with men.[32]

This existed more as an ideal than a practical reality. Women's roles remained segregated by gender and this was articulated in the wider press which continually noted the 'femaleness' of female announcers, characterizing the role according to relatable gender conventions that would not upset social norms.

The female announcer, then, represented on the one hand the progressive nature of the BBC and its alignment with economic and political equality. On the other hand, the female announcer pointed to a cultural and social conservatism whereby the role performed by, as well as the representation of, women would reflect qualities such as feminine passivity, discipline, charm and deference. The responses to the female announcer reflect this paradoxical position. She was not only signifier of the increasing role women would play in the public sphere (and which was both championed and demonized) but also codified in traditionally feminine terms. She represented the route through which issues of gender were contested and contained.

The female announcers of the BBC were equally subjected to gendered assumptions about performance and presentability. They were expected to connote glamour and class and their selection was suggested to stem as much from their social background as their professional one. Critic and journalist Kenneth Baily suggested that the women were appropriately upper-class,[33] implying that BBC Director of Television Gerald Cock 'would probably have said that for *savoir faire*, dignity of bearing and charm you could not beat the socialite strata of hard Mayfair training'.[34] The women, then, would function much like the television girls – feminine ideals who would act as objects of pleasure rather than as figures of authority. This dismissal of female agency meant that the BBC could maintain the outward appearance of progress, while perpetuating traditional gendered social hierarchies.

In its promotional broadcasting guide, the *Radio Times*, this emphasis on the aesthetic function of the female announcer continued. A feature on Cowell played upon her naivety in comparison to the demanding technical environment. She is described as a babe-in-the-woods:

> A slim figure in a white jacket, with a black skirt, her dark brown hair as attractive as her brown eyes … she is silhouetted against a background of grey curtains draping pale flats of scenery, as she faced a terrifying camera mounted on a trolley with pneumatic tyres and waited for zero hour.[35]

The appointment of the female announcer was managed in a way that reflected this paradox. The BBC had employed female radio announcers in 1934 with limited success and little popular appeal, and the decision to terminate their employment apparently resulted from the negative reaction from women listeners.[36] Echoing the concerns about female announcers in the United States, the female announcer was subject to a set of debates that evidenced caution and concern about women in mass media. The BBC managed this dilemma by integrating the role of announcer with that of hostess so that the announcer would not only lead on the delivery of programme talks but also undertake the more typically feminine responsibilities of greeting the guests before the programme transmission. The BBC tempered the potential disruptive force of an authoritative female announcer by assigning her duties of care. Bligh and Cowell were also both educated women of high social standing, with Bligh a descendent of Captain William Bligh of the Bounty fame.[37] Attention was directed to their background and skills which worked to guarantee their suitability for the role. 'Miss Bligh … has three years of stage and film experience. She speaks French fluently … Miss Cowell … has travelled extensively abroad and speaks French and German. She has considerable experience in mannequin work and has specialized in dress design and display.'[38] It was reported that they underwent a rigorous three-stage application process in order to ensure that they possessed 'the necessary qualifications of personality, tact and charm, had "photogenic" faces, suited to the television medium'.[39] The careful public relations management of the female announcer mitigated the potential hostile reaction to her. The BBC, along with the press, sought to cultivate a safe and inoffensive identity for the female announcer who would reflect an idealized mode of femininity and domesticity that would not upset the status quo (see Figure 5).[40]

Figure 5 BBC television announcers, from left to right, Jasmine Bligh, Elizabeth Cowell 1936, with permission of BBC Archives.

Perhaps because this image was rendered safe, Bligh and Cowell could be championed as evidence of women's growing presence in the mass media and participation in wider public affairs.

The female announcer, then, served as a prism through which debates about gender were enacted. Bligh and Cowell bore the weight of wider social expectations about what modern womanhood would mean. Their programmes were subject to equal measures of criticism and praise; the women were criticized both for their lack of dynamism and for their very presence on television. For example, where some pointed to the lack of screen time given to Bligh and Cowell, others justified this on account of women's lack of skill in the role of announcer.[41] A scathing review of their performance pointed to expectations of the inevitability of their underperformance:

> The newly-appointed television hostess-announcers, Miss Jasmine Bligh and Miss Elizabeth Cowell, were given little to say in a variety programme broadcast on Tuesday and one had no opportunity of forming an opinion of their 'microphone personalities'. In any case, one of their primary functions is to please the eye – as much as the ears – and it would be incorrect to regard their inclusion in this week's variety bill (at any rate from the listeners' point of view) as anything more than a tit-bit novelty. Neither was free from slight anxiety, nor had gained the confidence of the 'hello, folks' calibre which comes form [*sic*] experience. But naturally, nothing other was to be expected.[42]

The hostile reaction to the women served to reinforce masculine authority and redraw the boundaries of gendered professions whereby women were, by fact of gender, unsuitable for television presentations. The challenges posed by women's entry into the masculine cultural sphere were alleviated by the confirmation of their disruptive presence.

This gendered discourse was equally evident in internal communication in the BBC. Jasmine Bligh, in particular, was painted as a childish and incapable announcer who was troublesome. Director of Television Gerald Cock referred to her as a 'spoilt child' who 'needs (or expects) a bit of pampering'.[43] Cock was particularly scathing of Bligh at times and referred to one particular instance of Bligh's work as 'the worst I have yet heard by television; rushed, meaningless, not synchronising accurately with the artists before the camera and sounding ill-tempered.'[44] Cock pointed to the numerous complaints about Bligh's work and issued her a warning about her performance. Assistant Director of Television, R.A. Rendall, echoed Cock's views on Bligh's work by suggesting that he 'did not think she would ever make a big success of studio announcing.'[45] Such critical views of women's announcing sat alongside a competing view that framed the female announcers as examples of the professional capabilities of women in the labour force. Women would have professional roles and responsibilities but they would be scrutinized and held to extremely high standards. Women would introduce television to the public, but bear the responsibility for its failure should they be less than stellar in the performance of their duties.

Ultimately, and despite the criticism internal and external to the BBC, the BBC had female announcers in its first three years of experimental television broadcast. The use of women was perhaps a calculated risk since audiences remained low until the late 1940s. Thus, the use of women to showcase television was unlikely to have been perceived as having too many negative consequences. Where radio broadcasters were concerned with the presence of women announcers on the radio airwaves, television, in the first instance, seemed comparatively welcoming to the individual women employed as announcers. This openness was perhaps due to television's novelty and its status as peripheral, the 'juvenile' relation to radio broadcasting. Given the low viewership of BBC television, the risk of having female announcers may have seemed low and women were allowed access to the medium while this was the case. In addition, during these early years, television was, for much of the public, something to read about in magazines and newspapers. This allowed the BBC to experiment not only with television programmes but also with its use and employment of women in key positions. At the BBC, for example, a number of women held salaried positions. Murphy notes how waged and salaried positions were gendered. Advertised waged positions were more clearly defined as men's or women's work. Salaried positions made available to women were usually 'clustered in areas that were viewed as appropriate to their innate natures such as women's talks, School Broadcasting and *Children's Hour*'.[46] However, she also notes that at the BBC gendered 'demarcations were less clear cut' than the wider economy.[47]

This complex practice and discourse are evident in the schedule of female-produced programmes as well as the *Radio Times*' discourses of women at work at the BBC during the experimental years of television. The BBC carefully negotiated a space for women's roles, particularly in terms of balancing their professionalism with their domesticity and femininity. Women were defined and represented through discourses of production and reproduction, public and private, professional and domestic. By emphasizing employed women's femininity and domesticity, the BBC could manage the integration of women into typically male domains of production. For example, in a 13 November 1936 issue of *Radio Times*, this materialized in a television programme summary of 'women's interest' programme, *A Laundry Demonstration*.[48] In the programme synopsis, the speaker, Daisy Pain, was identified as authoritative in her skills as both a broadcaster and a homemaker. 'This is a practical demonstration by a practical woman. Mrs. Daisy Pain, as well as being an expert broadcaster, is an experienced laundress, for she has worked in and supervised her husband's laundry.'[49] While Pain was largely introduced through the prism of domestic labour, the summary also conferred professionalism (and indeed class identity) on her as someone capable of managing the household and as an expert.

In the first year of experimental television (1936), women undertook many such roles on television programmes, acting as authoritative demonstrators of domestic tasks such as cooking, ironing and washing. These sat alongside other demonstrations and talks on gardening, architecture and DIY, usually presented by men. However, although the BBC was somewhat open to representing the expertise of women in such tasks, the BBC's own television audience research indicated that two-thirds of the viewer comments that related to studio demonstrations and talks were negative and that 'disapproval concentrated largely upon the demonstrations of cooking, washing, ironing, etc., which were condemned as of little interest to those who could afford television sets'.[50] This was perhaps less a rejection of women in television as of the appropriateness of programme subjects to the audience. Those who could afford television sets were likely the same who could afford for washing and ironing to be carried out by paid domestic labourers. Perhaps in direct response to the criticisms of the programmes on domestic chores, the BBC ultimately steered away from such demonstrations and talks, and by 1937 there were few instructional programmes on domestic chores.

The BBC carefully constructed representations of female expertise on its television programmes. Where Bligh and Cowell were, at times, infantilized and patronized by viewers and by staff in the organization, educated and 'skilled' women were celebrated, particularly in experimental programmes about or aimed at women. The series *World of Women*, transmitted in 1937, championed women in the fields of arts and culture. Although the title of the series immediately gendered the content, it can also be understood as positioning women as key figures in the various fields featured on each programme. Aimed at an educated audience with cultural capital, the programme was advertised and promoted as appealing to a cultured viewer. The promotional material for the programme stressed women's authority and expertise on subjects and activities beyond those typically understood as feminine. The *Radio Times* synopsis for the premiere episode 'Setting a Play: Molly McArthur' (12 January 1937)[51] emphasized McArthur's experience in theatres and listed the productions she was involved in.[52] It also noted her studies abroad, thus foregrounding her authority and legitimizing her contributions to the programme. The *World of Women* summaries for each of the episodes pointed to the complex means by which female authority was discursively produced. Such authority was, at times, legitimated through social class, social titles, education and marriage, as well as through talent and skill. This framing of the female expert as a product of institutions of authority (as well as productive of authority in the 'world of women') safely introduced female authority to the audience. For example, women co-demonstrated on programmes featuring fencing, ju-jitsu, bridge and horse-riding.

Outside of *World of Women* other equally informative and educational talks were undertaken by women and without reference to gender. The programme *Expedition on a Bicycle*,[53] for example, was presented by Myfanwy Evans who recounted her recent tour of the Chilterns. Although, like other female talent, Evans' authority was framed through her broadcaster husband, her education and journalistic experience also served to qualify her to deliver a programme on the subject. This was perhaps equally significant given the historical association between feminist activism and the bicycle. Elsewhere, the *World of Women* episode 'Painting and Pottery' (5 February 1937)[54] introduced Dame Laura Knight who was, in the words of the *Radio Times* programme synopsis, 'in the first rank of English artists'.[55] However, the synopsis also mentioned her marriage to Harold Knight. Similarly, Lady Kennett, the expert for the episode on sculpture, was referred to as the widow of Captain Scott and wife of Lord Kennett. The summary then went on to note her own achievements.[56]

The synopses thus gave equal weight and value to the women's titles and marriage as to their expertise. While the references to social status were perhaps an effort to associate television with prestige, the means by which women were introduced in these *Radio Times* synopses also worked to associate them with social power and significance. And while these instances suggest that women gained social power and authority through the family patriarch, the BBC (through the *Radio Times*) also identified and promoted women's independent achievements and expertise. The *Radio Times* synopsis on the sculpture programme, for example, emphasized demonstrator Dora Clarke's success in gaining an educational scholarship.[57] Her work and research in Africa suggested ambition and experience. The synopsis of the *World of Women* episode featuring filmmaker Mary Field celebrated Field for her brilliant documentaries and her accomplishments in filmmaking.[58] The *Radio Times* featured an article by Field in which she outlined the processes of camera operation for nature filmmaking.[59] Alongside images from her films, an image of Field 'in the cutting room' worked to confirm her as central to the technical operation of filmmaking. She was pictured with a female assistant inspecting a reel of film. A caption described Field as 'one of the leading experts in the production of nature films'.[60] The *World of Women* series, then, could be considered as evidence of the BBC's role in generating a sense of normalcy about women in the professional and cultural spheres. In comparison to other programmes centred on 'women's interests', such as those on domestic responsibilities, the *World of Women* worked to include women in more general interest areas not defined by gendered responsibilities.

Such representations of female expertise were the result, no doubt, of the programme's producer, Mary Adams. Women such as Adams and assistant editor for *Picture Page* (1937–9) Joan Gilbert ensured that women were represented in front of and behind the camera. The television department was both new and quite small during the experimental years and this meant that women had opportunities to innovate and gain experience. The BBC was a progressive organization in many ways, but as elsewhere, women had less access to the professional sphere. In addition, as Murphy notes, there were both class and gender barriers at the BBC.[61] There were numerous barriers to entry and difficulties faced by women in the organization. Salaried women, such as Adams, were often steered towards areas deemed of particular interest to, or requiring the expertise of, women (women's talks, educational broadcasting, etc.), especially in post-war television.[62] In addition, the marriage bar, as Murphy notes, allowed the organization to selectively keep or lose female members of staff as it saw appropriate.[63] As Kathryn Terkanian points out, those in the BBC could often circumvent the marriage bar by employing women on short-term contract, which, of course, resulted in precarious employment and limited opportunities for promotion.[64] Murphy's extensive research of the employment practices of the BBC indicates that, despite these limitations on women's employment, the organization was eager to showcase its progressive practices. Murphy points to the publication of the internal staff periodical, *Ariel*, published from 1936, that included women among its news updates.[65]

Similarly, the *Radio Times* television supplements of the years 1936 to 1937 drew external, public attention to the role of female employees in the sphere of television, including not only the announcers Cowell and Bligh but also Head of Make Up and Wardrobe, Mary Allan, switchboard operator, Joan Miller and producer, Mary Adams. Given that television production during the experimental years was limited and the television department fairly small, it seems that there was a significant presence of women both in front of and behind the camera. And given that television was in its infancy, these women were often pioneers in the field of television. Many broadcasting and entertainment periodicals noted the work of these women and drew attention to their roles in the formative development of television and demonstrated a distinct confidence in the women who played a role in the development of the medium. Thus, while one discourse worked to frame women as the objects of pleasure on the screens of television, other discourses worked to frame women as productive and creative contributors to emerging television.

Through the *Radio Times*, the BBC set about generating discourses of women's professionalism and skill in television. Mary Allan offers such an example. As the first Head of Make Up and Wardrobe, Allan was responsible for developing techniques and processes especially for the new medium.[66] She joined the BBC television department in 1936 having worked in similar roles in film production prior to that. Since Allan was married, her position needed to be considered at a marriage tribunal, where it was ultimately decided that she should be employed on the basis of her skill and professionalism.[67] Allan worked across a range of programming during the years 1936 to 1939, including the aforementioned *Picture Page*,[68] *Julius Caesar* (1938)[69] and *R.U.R.* (1938).[70] Although Allan was qualified and experienced in make-up for film and stage, television necessitated a new set of practices. In a *Radio Times* feature on 2 April 1937, Allan noted that her work required collaboration with engineers and technical departments such as lighting and camera.[71] She noted that make-up practices in television were distinct from film and she detailed some of the problem-solving skills she put to use on the job. Her account suggested her own self-confidence as well as the confidence the BBC had in showcasing its female television production crew.

Equally, many other reports of Allan's employment, her contributions to television and her role as make-up artist stressed her professionalism and practical competence. While an October 1936 issue of the *Radio Times* referred to Allan in much the same gendered terms as the female announcers – 'attractively slim, medium height, deep-set eyes'[72] – other magazines and periodicals focused on her contributions to the development of television programming and on her hard work. Upon the announcement of her employment, for example, a 1936 issue of *Wireless World* (a technical broadcasting periodical) noted that her appointment was 'of considerable technical interest for the task of securing a correct balance of shade at the television transmitter is tantamount to that of balancing musical instruments in a sound broadcast'.[73] A later issue of *Wireless World* noted her expertise and championed her role in the development of early television. 'Mary Allan, the make-up expert, is among the heroines of these early days of television'.[74] Elsewhere, magazines and television reports drew attention to Allan's education, her prior work experience, her skills and expertise, and her hard work. The periodical *Television and Short-Wave World* noted that Allan was 'one of the busiest people at the Palace … ' who is 'equal to anything'.[75] The article elsewhere remarked upon her speed, demonstrated in her 'record-breaking' ability to make up thirty faces in thirty minutes.[76] The *Sunday Times* made much of her twelve-hour working day and detailed her education: 'She spent a year at a London hospital studying anatomy and the structure of the face'.[77]

Finally, TV critic Kenneth Baily's reflection on the 12 May 1937 television production of the Coronation emphasized Allan's capabilities and professionalism. He detailed Allan's role as such:

> Whatever happened in the hectic life which electrified those studios this was the one woman who maintained calm and unruffled good

humour, and by it saved many trying situations … Her job, obviously, was to paint their faces; but while she did this they discovered that she also calmed their nerves. To her craft she added an intuition and psychology which were remarkable.[78]

In terms, then, of how female television crew were publicly presented, discussed and understood, it would seem that some effort was made to normalize and celebrate women's work in television production. As technical periodicals, *Wireless World* and *Television and Short Wave World* accounted for the significance of Allan's work in contributing to television presentations. The *Radio Times*, having an obvious vested interest in promoting the professionalism of its television personnel, equally foregrounded the contributions of Allan to television's development at the BBC. While this is in no way representative of the experience of Allan as an employee at the BBC nor of conditions under which women operated, it does point to some of the ways that women participated in the formation of experimental television at the BBC and of the way in which this was part of a more public discourse about how the new medium of television would be representative of more progressive work cultures at the BBC.

In the experimental years of television in 1930's Britain, the presence of women was noteworthy. The BBC had already instituted policies aimed at generating a more inclusive work culture (even though the extent of that inclusion was limited to certain positions). Although the BBC television department inevitably drew from the organizational model and structure of radio broadcasting, as an innovative department that was developing a new form, it was somewhat open to experimentation. This allowed the women who were part of early television to contribute to the form, style and production practices of the medium. While this was ultimately framed within a distinctly patriarchal organization, it was nonetheless significant that women could innovate with television formats and that these innovations were acknowledged and publicized by the BBC. In addition, while radio broadcasting had by the mid-1930s developed a programming model that addressed female listeners largely as homemakers and mothers with domestic responsibilities in the private sphere,[79] television broadcasting during the late 1930s both represented and addressed women as engaged in more public activities such as sports, sciences and arts.

This was, in no small part, the result of the contributions of Mary Adams (see Figure 6), the BBC's first female producer, who transferred from Radio Talks to the television department in January 1937.[80] Adams, a biologist who had lectured in Cambridge, had already established herself as a prolific radio producer of programmes in the fields of science, politics and society. Adams's relationship with the BBC began with a radio series *Problems of Heredity*[81] in 1928 and she was formally employed in 1930. She had initially unsuccessfully applied for the role of Director of Talks in radio, and so the offer of a role in television may have seemed as a rejection of her ability.[82] Nevertheless, the volume and range of programmes produced by her demonstrated her skill and diversity, even outside of her previous areas of expertise. As a television producer she, like her male counterparts, was less visible in the public eye than female announcers like Bligh and Cowell, but her presentations were attributed to her in the *Radio Times* television supplements (as 'presented by Mary Adams').

Among her programme's topics were architecture, the history of fashion, art, gardening and home affairs. As with her practice in radio broadcasting, Adams was capable of leveraging her network of colleagues and associates to entice a number of well-known public figures to television. Adams produced the programme *World of Women*, which featured artists such as Mary Field, Dora Clarke and Dame Laura Knight. She was also responsible for *Clothes-line* (1937),[83] *Friends from the Zoo* (1937),[84] *Architecture* (1937),[85] *The Future of Television* (1937),[86] *Clothes through the Centuries* (1938),[87] *Artists and Their Work* (1938),[88] *Spelling Bee* (1938)[89] and *Sight and Sound* (1939).[90] Given the small size of the television department (as well as its separate location at Alexandra Palace), Adams and the other television producers had relative leeway in developing programmes and innovating with the form. As with other producers such as George More O'Ferrall,[91] Cecil Lewis[92] and Cecil Madden,[93] Adams was responsible for one particular

Figure 6 Mary Adams, director of experimental television 1936–9, with permission of the BBC Archives.

format: Talks. Where 1937 featured a range of formats – including drama, variety and talks – fairly consistently across the schedule, by 1939 variety and fictional forms of television had increased more so than Talks.[94] Nonetheless, Adams's contributions to the development of television programmes during these years indicate that experimental television production was less gender segregated than it would be in later years.

An important trailblazer in television, Adams's career in broadcasting extended from radio to pre- and post-war television. Adams, for example, initiated, and worked in, the Women's Programming Unit. She also produced and advocated for educational and children's television.[95] However, while the experimental years of television afforded more opportunities for creativity and innovation within limitations, in the post-war years BBC television seemingly became far more bureaucratic. This was reflected in Adams's encounters with senior members of the BBC who were not inclined to resource Adams's department. This is evidenced in the experience that Adams had as Head of Television Talks in the late 1940s whereby her persistent complaints about the under-resourcing of the department fell on deaf ears.[96] A memo from Director of Spoken Word, George Barnes, to the Controller of Television, Norman Collins, pointed out that Adams and the other Talks producers were expected to undertake duties that the Drama producers were not.[97] Despite the support of Barnes, Collins blamed Adams for the poor organization of the department and questioned her leadership skills.[98]

Nonetheless, during these experimental years, Adams's contributions to television were clear in the volume of output she was responsible for. Initially focused on science radio talks, she moved on to become one of the first television producers of a range of programmes, the subjects of which included architecture, home affairs, art, gardening, history and science. Adams's role in the production of so many programmes and forms is reflective of the small department that made up television, but what is clear is that, as a female producer, Adams had significant presence in the delivery of programming. Even more noteworthy was the subject matter of the programmes in her early years with experimental television, which did not reflect gendered patterns of programming that had emerged in radio by this time and would later do so in television. Adams's productions were more defined by educational and cultural fare than by conservatively gendered-defined interests.

Through the *Radio Times* the BBC – at least in the 1930s – worked to emphasize her professionalism and experience as well as her competence in developing and producing general interest programmes. Throughout the years of 1937 to 1939, reference was made to her ongoing work in creating and producing a range of different shows and in illustrating her tasks as producer. For example, it announced her employment as television producer in the British press and foregrounded her prior work in radio as well as her specialism in science. This was re-circulated in national and regional newspapers:

> Mrs. Mary Adams, who has been attached to the talks department at Broadcasting House, mainly in an advisory capacity, has been transferred temporarily to the Television Department where she will plan talks features. Mrs. Adams was a Lecturer in Biology at Cambridge before taking up broadcasting ... Mrs. Adams, who is the wife of Mr. Vyvyan Adams, M.P. for West Leeds, has herself given radio talks and was concerned with an inquiry instituted by the London School of Economics into the changes which had taken place in family life during the last two or three generations. She gave radio talks on the subject.[99]

Elsewhere, reference was made to her professionalism in carrying out her role as producer in terms of employing others, in researching and developing programmes and in terms of technical processes. The *Nottingham Journal*, for example, noted that she was considerate when interviewing others for roles as announcers, referring to her as 'a good producer' who was 'human' and whose radio broadcasts would be remembered.[100] The *Radio Times* made reference to the technical responsibilities of Adams when accounting for the use of voice-over narration in the prospective programme *They Looked Like This*: 'without a script faithfully adhered to by the speaker, Talks Producer Mary Adams would have an impossible task at her desk in the control room in managing "fades" at the right moment'.[101]

More numerous were the *Radio Times* updates on Adams's influence on television where she was often noted as being actively engaged in the production of new programmes. It claimed that she had authority over the entire talks schedules in 1938 and was wholly responsible for the content produced for Talks. 'Producer Mary Adams tells me that her policy will be to continue series that have been successful, eliminate those that have not been, and start those that are likely to be.'[102] She was presented as someone who was assertive, energetic and determined in undertaking new programmes. For a programme on Soho café culture, Adams researched and sourced a range of contributors and subjects for a studio-based restaurant set-up:

> All this for a television documentary on Soho, to be produced by Mary Adams on April 19. For several weeks she has wandered round this most cosmopolitan part of London, gathering material. She is determined only genuine Soho characters will take part – a noble resolve, for the many professional actors who frequent the district must be very tempting.[103]

If any trace of criticism or hesitancy was evident in the *Radio Times* features on Adams, it was perhaps that her programmes might be perceived as too serious or elitist rather than a perceived issue with her gender. For example, effort was made to stress the appeal of her programmes beyond being purely informative. In promoting new programmes on talks by museum and art gallery directors, the *Radio Times* was keen to frame the programme as enjoyable as well as informative: 'With the *News Maps* and *Guest Night* series, television has shown that where instruction comes in entertainment does not necessarily go out.'[104] Another issue joked that viewers might not understand the poetic origins of one of Adams's programme titles, *Rough Island Story* (1939)[105]: 'I understand that visitors to viewers' homes when *Rough Island Story* is being televised have been causing a great deal of embarrassment. Another time, if they ask what is the origin of the quotation that gives the title to Mary Adams's popular series, say quickly "Tennyson, of course".'[106] Even here, the *Radio Times* acknowledged the popularity of the programme, even if the title was somewhat ridiculed. The reputation that Adams had (at least, according to *Radio Times* publications) was of a talented programme-maker who produced informative and entertaining programmes.

While Adams went on to have a prolific career in post-war television and produced many women's programmes, it is nonetheless important to account for her place – and the place of others such as Allan, Bligh and Cowell – in the earliest years of British television. The opportunities they had, the roles they played and the barriers they faced tell a story about women's experience in emerging media industries. By the 1950s, BBC television was firmly established and, with this, came a sophisticated and complex organizational structure. This was less the case in the 1930s when television was still finding its feet. As a medium on the periphery of the broadcast industry, there were perhaps fewer barriers to entry for women in early television production. During the experimental years, production practices and roles had yet to be defined and television studio production was impromptu and exploratory. Television production was not funded a great deal and the BBC was not entirely optimistic about the future of television. With low budgets and low public attention, those involved in television had more freedom to determine the form they thought television could take. This freedom also took the form of less hierarchical and less rigidly defined employment. Women who worked in early television were able to develop and demonstrate their creativity and professionalism.

From the late 1940s, the demand for television grew in Britain and this resulted in the professionalization of the television industry. In Britain, 'the expansion of television broadcasting [during the 1950s and 1960s] both required and promoted the expansion of bureaucratic and capitalist *rationality*'.[107] For Turnock 'Rationalization *in* television brought about an increasingly streamlined, efficient and cost-effective mode of production. With the expansion of broadcasting, it led to both *professionalization* and *industrialization* of television broadcasting'.[108] This resulted in a gendered division of labour that relied on masculine control over the fields of television production and female presence mainly in administrative and production roles that were deemed appropriate for women.[109] While women undoubtedly excelled in such roles (as was the case of Adams), they nonetheless found themselves in a post-war organization that had many more gatekeepers. A similar pattern was evidenced in early US television which is the subject of the next chapter. In the United States, like in Britain, women were part of early television and worked across a wide number of roles in productions. However, as with histories of British television, women's roles in US television histories have largely been discussed in relation to the 1950s onwards.

4

Women in early US television

In US experimental and early commercial television of the late 1930s and early 1940s women, as in Britain, worked in a variety of roles in production. Indeed, many women were part of the transition to commercial television in the 1940s. Commercial television began in 1941, although the capacity for television to develop was hampered by a number of factors related to the war, including a reduced production capacity as a result of manufacturing restrictions and the enlistment of male workers in the military.[1] The latter might have had an impact on overall television; however, it facilitated the entry of many women into the television industry. As with British experimental television, many women did find work in the novelty medium and found less restriction on job roles than they might have had in radio. Although it is difficult to estimate the numbers of women (or men) working in television during the 1930s and early 1940s, it is possible to see how women's television production work was discursively produced by the trade and popular press during this period.[2] As women began to work in early television in the late 1930s, they were often featured in press reports that associated television, through these women, with modernity and glamour.

The case studies discussed here include those of announcer Betty Goodwin, and directors Thelma A. Prescott and Frances Buss, who demonstrate the ways in which women's television careers were structured and defined in the press. Also discussed is one-time television producer Helen Sioussat who cultivated a career persona and managed the question of women's professionalism in broadcasting. These women may not have the lasting legacy of television icons such as Lucille Ball, Gertrude Berg and Ida Lupino; however, they demonstrate the range of roles undertaken and evidence wider participation of women beyond that already part of the television canon. Finally, I examine how the Second World War opened up opportunities in both radio and television production. Women benefitted from the general acceptance of their employment during this time as they were perceived to be supporting the war effort. NBC, in particular and, through its staff publications, championed women's support of the war effort and of their custodianship of broadcasting during this time.[3] However, as Ruth Milkman notes, 'the ideological definition of women's war work explicitly included the provision that they would gracefully withdraw from their "men's jobs" when the war ended and the rightful owners returned'.[4] Women's work was, according to this view, tolerated but not fully accepted. This was particularly evident in the case of Chicago station WBKB's Women's Auxiliary Television Technical Staff who took up the roles made vacant by men who had enlisted. Initially celebrated as the first 'all-woman' television station, these same women found less space for themselves in the post-war years. Not all women experienced such overt marginalization; however, as I argue throughout this chapter, the trade and public press served an important function in gatekeeping women's access to, and in negotiating the roles they might play in, the emerging television industry.

While histories of women's work in early British television have been fairly well mapped and continue to emerge, proportionately less attention has been paid to women's work in US television during the 1930s and 1940s. Despite the numerous institutional, technical and political histories of US broadcasting, there has been little centring of women's contributions to its development.[5] Where women are referenced, it is largely in the context of their role as television's audience or their performance on television. While it may be the case that fewer women than men worked in the early television industry, the absence of reference to them naturalizes television as a male domain.

Scholarly work has been done to account for the histories of women in early radio and, indeed, some women such as Gertrude Berg were able to transition from radio into television. A survey of the scholarship on radio also reveals certain patterns that were evident in television: opportunities available to women in the earliest years of the medium,[6] subsequent restrictions on women's roles as announcers,[7] the feminization of women's programmes and productions,[8] and the role that women played in the development of radio formats and genres.[9] Broadcasting scholar Michele Hilmes has made the case that women's work on daytime radio was highly innovative and creative. She argues that despite the gendering of broadcast hours and programmes, female programme-makers such as Irna Phillips and Mary Margaret McBride dealt with important female issues and concerns.[10] Donna Halper's *Invisible Women: A Social History of Women in American Broadcasting* profiles the work of female broadcasters between the 1920s and 1970s.[11] She notes how women such as Helen Wylie of KPQ in Washington began to manage radio stations in the place of their husbands who had joined the war effort.[12] However, as Halper suggests, a programme of 're-domestication' ensued following the war and many women working in radio (Wylie included) left their jobs. This pattern is reflected in the experiences of women in television too.

Among the accounts of those women's experiences are archivist Cary O'Dell's two books on women in television, which also undertake a revision of the 'great man' histories that tend to dominate studies of early television. In *Women Pioneers in Television: Biographies of Fifteen Industry Leaders*,[13] O'Dell identifies those women who, while never running or owning a network, made contributions 'in areas and genres where few, if any, men had yet succeeded or even made attempts'.[14] Women, in O'Dell's narrative, become key figures in the formation of the television industry and of television programmes. O'Dell constructs a model of female authorship for figures like *I Love Lucy*[15] producer and actress Lucille Ball and *As the World Turns*[16] writer Irna Phillips. O'Dell's book *June Cleaver Was a Feminist!: Reconsidering the Female Characters of Early Television*[17] argues for a more favourable analysis of female representations in 1950s television shows since women often acted as writers, producers, directors and performers on their own programmes. Erin Hill's *Never Done: A History of Women's Work in Media Production* traces women's work across the emerging media industries during the early twentieth century to identify the spectrum of roles that women undertook and to foreground the significance of roles typically excluded from dominant institutional histories.[18] Elana Levine's research on the soap opera is equally recuperative, in that she argues that the soap opera is less a 'ghettoized' women's genre that lies on the periphery of American television and more a central and formative feature of US cultural history.[19]

An area that has marginally more attention is television news and journalism. Although women were largely restricted from presenting television news for its first few decades, much has been written about women's inventiveness in and contribution towards news and current affairs programmes as well as the barriers to entry to careers in broadcast journalism. Mike Conway, for example, notes that during the 1940s and while women were being hired to work in radio and television newsrooms in technical and operational roles, they were prevented from reporting the news.[20] He cites the case of Ruth Ashton, who was hired by CBS television in 1948 working on election coverage, was then moved to religious programmes and left CBS when she was not allowed to report the news.[21] As with radio, it was taken as common sense at CBS that audiences would not respond well to female announcers. David Hosley and Gayle Yamada also note similar resistance to female news announcers in the early years of broadcasting.[22] They point to the indefatigable efforts of Pauline Frederick to get a job at CBS and NBC but who contended with an industry that didn't have a need for more women.[23]

Others have produced more focused studies of individual females, programmes or stations and find a diverse range of roles undertaken by women under various conditions. Stacy Spaulding makes a similar case as Levine about the ghettoization of female news broadcaster Lisa Sergio, arguing that her legacy should be as a pioneering radio news broadcaster rather than as a women's news broadcaster (as had been claimed).[24] David Ozmun's study of early television news concentrates on the work of Natalie Jones for NBC programmes in the year 1952.[25] Ozmun points to an issue common to historians of women in television: the lack of NBC documentation recording Jones' employment. Jones worked with her husband and, since RCA (NBC's parent company) disallowed relatives to work together, Jones' was referred to not as a reporter for NBC but as a 'contact' or 'liaison'. Although she worked as interviewer, photographer and sound recordist for NBC productions in Europe, her work was not formally recognized when, for example, her husband won awards for their reporting work. This collection of literature offers a number of biographies, oral histories and institutional narratives that frame the work and careers of women working in the emergent television industry.

Ultimately, what is revealed in the literature on women in early US television is a rich but elusive and contested history that is found not in the dominant narratives of the formative years of television but on the periphery or scattered across archives, diaries, trade magazines, oral histories and libraries. Where those dominant histories are readily available to scholars and to the public, the histories told from the margins and through the experiences of women must be treated somewhat like a puzzle to be solved. Time and again, the literature emphasizes how hidden these histories are and how dispersed they are across any number of sites. Unlike the 'great men' of television, few women's stories – especially those working at the coalface – were deemed historically significant enough to draw attention to, exhibit and put at the centre of narrative production of history. In this sense, women *became* absent rather than *were* absent from television history. This has resulted in a popular history that tells a story of a masculine industry concerned with feminine reception and consumption. It has also resulted in an impression of a neat chronology that finds few women present in early television, with more and more entering the industry through the years.

While the literature largely traces women's participation in television from the post-war years, this chapter seeks to extend further back, to the experimental years of television during the 1930s and 1940s. The women discussed throughout this chapter were not the 'great women' pioneers who have become household names. Instead, I focus on stories of women in early television that evidence a wide range of experiences and interactions with the industry. This includes women who were successful in maintaining careers, women who were not able to sustain careers, women who worked briefly in television before leaving, as well as those who moved on to other media roles. Included here are women who worked in 'above the line' and 'below the line' work, although I recognize that, especially in the early years of television, these were certainly not hard and fast categories. The women discussed were all part of the press and promotional discourse produced in the early years of television.

This chapter contributes to the histories of women in television by tracing the way in which the trade and popular press create a discourse of the female television worker that negotiated what was at that time a conflict between women's social role as domestic and feminine and their increasing participation in the emerging television industry. I turn to hobbyist, fan, trade and popular periodicals and magazines of the era to identify how women's roles were framed and structured in emerging industrial literature. Trade press reports on women's work helped to normalize women's work in television while, at the same time, defining and limiting what their roles might be. As insider publications, these acted as gatekeepers in terms of what was permissible or not for women to undertake. Employee-facing publications like *NBC Transmitter* offered regular features and updates on female staff and the roles they undertook. Equally, widely read publications such as *Broadcasting*, *Billboard* and *Variety* proposed norms for industry practices, conditions and trends and spoke 'for industries' and 'not just about them'.[26] These trade stories, which reveal the industry's discourses for and about itself, are evidence of how the industry's workers are conceptualized and narrated through trade press.[27] The conversation about women's work was an important one for the industry to have in terms of establishing recognizable roles for women, for example, as stars, models, actors and consumers (roles already in practice in the radio industry). In addition, these publications used stories of women's television work to signal an industry that was modern, if not progressive, with women ready and willing to turn their hand to 'masculine' work such as producing or directing programmes. Finally, during the war years, women were praised and celebrated for stepping into television roles vacated temporarily by men.[28] This overall narrative of openness was less apparent once the war was over and as the television networks began to expand and the industry professionalized.

The television landscape that these women entered was a widely dispersed and complex one. Unlike Britain, which had regulated for one provider in the BBC, the United States had a number of providers in the early years. As with Britain, television was, during the pre-war years, largely an experimental affair. The Federal Radio Commission (FRC) issued experimental television licences to companies and individuals from 1928, many of which were based in major cities such as New York, Chicago, Los Angeles and Washington, D.C.[29] Until technical standards were agreed, the Federal Communications Commission (FCC) refrained from issuing any commercial television licences. Innovators such as Charles F. Jenkins and Phil T. Farnsworth were granted licences to experiment with mechanical and electronic television technologies respectively. RCA, CBS, AT&T and GE also held experimental licences for technology and/or programme development.[30] In addition, individual radio stations were also experimenting with television. RCA dominated television research and was able to continue with television experimentation and testing through the Depression years.[31] However, by the mid-1930s there was growing competition from companies such as Philco Radio and Television Company and Allen B. DuMont television. CBS was also cautiously eyeing the television market.[32] There was, by the late 1930s, quite an active landscape of experimental television production and broadcast, with companies such as DuMont entering the field of set manufacturing. RCA was eager to pursue commercial television broadcasting and hoped to use the 1939 New York World's Fair as a platform to launch television in the public. For many, the 1939 World's Fair would be the first contact they would have with television. Here, television was introduced through a scheduled broadcast by NBC and attendees could get to see the first publicly demonstrated television images as well as the television sets that were priced beyond most people's means.[33] NBC offered this regular schedule of broadcasts in 1939 in order to demonstrate that a television service was possible.[34]

By 1940, the FCC was persuaded enough to allow limited commercial television for stations undertaking programme experimentation (but not technical experimentation since standards were not yet agreed).[35] NBC and CBS were the first to go on air in New York and offered advertising-supported broadcasts. CBS's commercial television station WCBW in New York had eight hours of programming a day until the United States entered the war, after which it dropped to four hours.[36] While television was widely publicized, broadcasters remained cautious about the new medium. So the lack of a significant audience worked in their favour, since they could experiment without the risk of deterring large numbers of people. By 1 July 1941, the FCC proposed the full commercialization of television with a minimum of fifteen hours of broadcast per week set as a minimum standard for stations. The schedules for broadcasters such as NBC and CBS were fairly similar and featured test patterns, sports events, variety shows and short news broadcasts.

The US entry into war stalled much progress on the growth of television with only seven stations continuing to broadcast.[37] A decision by the War Production Board in 1942 that 'materials used to manufacture television receivers and transmitters were essential to the war effort' meant that television growth stalled.[38] With no sets being manufactured, there were little incentive for advertisers to sponsor programming and, consequently, little incentive to continue developing programmes. However, some stations continued to experiment with the form and, with many men leaving the broadcast industry to join the war effort, this created opportunities for women to step into production roles. By 1945, the FCC began issuing television licences again.[39] It was after this point that the television networks emerged. Since NBC and CBS had already established radio networks and, therefore, had the knowledge and experience of broadcasting, they easily transitioned into television network service. NBC emerged as an early network leader. In 1945, it began to expand into television through its station WNBT, which provided programmes to affiliate television stations once interconnectedness of television broadcasts became available across the country. By 1947, NBC had established a network on the East Coast that included stations WRBG in Schenectady, WPTZ in Philadelphia and WNBW in Washington, D.C.[40] CBS was slower to establish a network since it had been holding out for new television technical standards that would aid its development of colour television. But these never materialized, and only in 1947 did CBS commence its network drive. By the end of 1949 it had one station, WCBS in New York, and twenty-seven affiliates, making up 27.6 per cent of the total market.[41]

However, in comparison to the slow development of television in the preceding decades, television's expansion accelerated quickly in the late 1940s. Television receiver purchases increased, the networks expanded and advertising revenue increased.[42] This is the institutional context that frames my discussion of women and television production work during the mid-1930s and late 1940s, and the media discourses that functioned to determine how women might participate in the emerging television industry. By the time the US television industry was established in the 1950s, women were addressed and discursively produced largely as consumers and viewers. During the experimental and early years under discussion, however, women's work in production formed part of the discourse of television that materialized in the trade and popular press.

This discourse of inclusivity and opportunity was evident in the various representations of television by broadcasters and in the press as a medium that promised open doors in comparison to the closing doors that was becoming the norm in radio broadcasting. However, the reality was that any such opportunities were limited. Indeed, the shift of the television industry from being open to aspiring women to being closed followed the trend that had been established in radio. After all, the earliest years of radio held the promise of access not only for women but for minority groups, as well as more localized communities and special interest groups. In the early 1920s, for example, African Americans were producing broadcasts through local clubs and organizations.[43] By 1924, Hispanic programmes were produced and distributed on white-owned radio stations.[44] Throughout the 1920s, ethnic and foreign language programmes were created by brokers and broadcast on local channels albeit in the cheaper slots in the day's schedule. In 1928, Chicago station WSBC became the venue for the one of the first African American-crewed and produced radio programmes[45] hosted by media figure Jack L. Cooper, called *The All-Negro Hour* (1929–35).[46] Women such as Eunice Randall and Ida McNeil were able to begin broadcasting careers by working in local and experimental stations as managers or announcers. Ultimately, though, radio of the 1920s came to reflect the social atmosphere of the day in demonstrating an overt hostility to diversity, inclusivity and multiculturalism and a favouring of social and cultural homogeneity.[47]

A similar trajectory is evident in television whereby early moments of inclusivity remained largely that: rare instances of women's ability in and aptitude for television broadcasting. This is exemplified somewhat in some of NBC's early television demonstrations and experiments that involved women. Here women were both present in and producers of some of the formative television broadcasts and programmes. Women such as Thelma A. Prescott, Chicago station WBKB's female technical staff and performer Ethel Waters took part in some of the earliest television broadcasts. Although they were used for their 'feminine appeal', their star status and their crowd-pleasing performances, their early presence and visibility nonetheless suggested an industry that would have the space and enthusiasm for diversity. Ethel Waters, for example, was one of the first African American women to appear on television. Capitalizing on her huge popularity as an actress who had starred in the Broadway play *Mamba's Daughter*, NBC broadcast *The Ethel Waters Show*[48] on 14 June 1939 on its experimental station WX2BS. The programme featured performances of segments from the play along with actresses Fredi Washington and Georgette Harvey. It is notable that during the same week and in the same city of New York, NBC was also cultivating the female television personality by holding a competition for the title of 'television girl' at the World's Fair. Both *The Ethel Waters Show* and the competition were aired on WX2BS.[49]

The impression might have seemed that television was inviting not only of white women but of African American and minority women too. A glance at the television schedule for that would have emphasized women's place on television. This was not the case, however. The invitation extended to women was conditional: firstly, on their racial and ethnic uniformity, with white women gaining more access than minority women; secondly, on women having a largely aesthetic function on television or in the promotion of television.

This trend was evident in the discourses about the female television announcer. Women's roles as early television personalities and announcers followed a pattern similar to that of the television girls of the early 1930s as well as that of Jasmine Bligh and Elizabeth Cowell of the BBC in Britain. In other words, their roles were somewhat ornamental and functioned to demonstrate the unique selling point of television, which was its visuality. NBC, in particular, was active in using the women to generate publicity about television and capitalized on the beauty and style of female television personalities such as Dorothy Knapp, Natalie Towers and Betty Goodwin. Goodwin, in particular, was recruited by RCA to introduce NBC's television programmes during experimental television demonstrations. Goodwin was an experienced broadcaster having worked for a small radio station, for newspapers and then for NBC at its Press Office and later for NBC radio from 1935 as its Fashion Editor and occasional writer for media publications.[50] Each of these roles provided Goodwin with the tools to work on RCA's first public broadcast at the Rockefeller Center on 6 November 1936, not only during the broadcast itself but on the associated publicity for the broadcast. During the broadcast, Goodwin commenced the television demonstration and introduced singers such as Hildegarde. Goodwin also hosted the fashion parade that formed part of the television programme.[51] Given Goodwin's experience in the NBC Press Office and on radio, she was able to provide a level of professionalism necessary for convincing the audience that television was a worthy and serious medium. However, Goodwin's image was used to demonstrate the clarity and quality of the televised image and was commodified as part of the emerging discourse of television culture. This is evident in how photographs of Goodwin's face on the screen were used to promote television to the public. A September issue of *Listen* magazine, for example, featured a photograph of a television screen image with a close-up of Goodwin smiling. The photograph caption used Goodwin's image to convince the reader that television would be aesthetically pleasing and that it would be a family activity, even though it was still a long way off.[52] Elsewhere, broadcaster Margaret Cuthbert, likewise, noted the importance of appearances when she championed Goodwin's television work: 'along with all [Goodwin's] qualifications, Betty is lovely to look at'.[53]

The use of the female announcer also drew from the established discourse and function of the Hollywood star, particularly in relation to the association between women and consumption in the consumer and marketing industries. The promotion of the female announcer worked to generate the television industry's version of stardom – the personality – and, through this, to nurture a notion of television as a commodity. As a representative of the new medium, the female television announcer formed part of the marketing and promotion of television culture. In effect, her role was not so much to be the communicator, the active agent of television, but to symbolically represent television as the latest consumer good. This is evident in the publicity and promotional literature that focused less on what communicative skills female announcers should have but on how they would appear to the audience. The emphasis on women's appearance dominated early press discourse of television's arrival and women's role in it. A March 1937 issue of *Modern Mechanix* stated that 'television announcers must not only speak well, but be attractive, too. Here is Miss Betty Goodwin, NBC's television announcer.'[54] Even Goodwin herself insisted upon the necessity of beauty in an interview with *Radio Daily* in October 1937:

> One of the few positive statements that can be made about television at this stage of the game is that it's sure to spruce up us women. From Hoboken to Walla Walla we'll be seeing today's fashions today. And if

it's the television camera we're facing, instead of the television screen, then the need for being clothes-conscious and figure-conscious will be even more acute.[55]

This focus on the relationship between television and consumerism as well as women's role within this was clear in a 1 December 1937 *Broadcasting* feature on fashion programmes presented by Goodwin.[56] The feature detailed experiments carried out to test the possibility of developing television techniques for showcasing jewellery and accessories in close-up, as well as make-up suitable for televising, emphasizing the association between women, consumerism and aesthetics.[57] This role of the announcer, then, was quite a prominent and visible one; yet, it inferred that women were the products of or on television rather than the producers of it.

Elsewhere, the trade press speculated about whether women might be able to enter television as its producers rather than announcers or consumers. As television began to emerge as a potential extension of or even competitor to radio broadcasting, women were occasionally imagined to have some role to play in the production of television. During the late 1930s and as far as the end of the war, magazines and journals enthusiastically featured the variety of jobs women could undertake in television production, suggesting that the industry was discursively imagining itself as a place open to women and ready for their participation. For example, a July 1939 issue of *Broadcasting* had a feature on 'Women's Place in Radio Advertising' that included references to the emergent television industry.[58] Offering advice for 'Directing a Woman's Program' and including a range of roles such as commercial manager, station manager, secretary and assistant, it outlined the opportunities that might materialize for women in television, largely within advertising:

Many women are anticipating some part in television advertising. It is believed that the first use of this new medium will be in the more extended demonstration of products; especially foods, home equipment, cars, and cosmetics. There are three ways in which you might participate:

1. Visually, as commentator or demonstrator.
2. By coming in as a voice, off-stage, delivering the commercial.
3. By writing the commercials and stepping into the visual area when you deliver them.[59]

At the point of emergence, then, there was much excitement within the broadcasting trade press about women's participation in the production of television.

Elsewhere, there was certainty that the television industry would remain open to women after war, so long as women stayed in their gendered corner. A feature in *Televisor* during 1945 noted that:

> While many of the jobs filled by women today have resulted from wartime man power shortages, there is little question in the minds of many television executives that women will continue to fill important jobs in the post-war period, especially as writers, producers and directors of women's fashion shows, children's shows, in fact any show with a woman's angle; as art directors, costumers, hairdressers, and make-up artists; as production assistants, talent agents, and in publicity, advertising and promotion … In the meantime, women today are showing what they can do in television – from station managers to 'cable engineers' – and being smartly videogenic at their jobs.[60]

This was, indeed, somewhat representative of what ultimately became of many women who worked in television during these years as it alluded to women's requirement to make way for returning servicemen. In other words, during a period of 'talent' austerity, women were considered quite capable of fulfilling those roles typically carried out by men such as camera operator, director and producer. However, as the *Televisor* feature suggests, once the war was over, women's work would be framed in relation to their gender. This evidences the paradoxical situation in which women were at once recognized as productive, capable and crucial to television while, at the same time, they were expected to give up that work for men returning from the war. The *Televisor* feature subtly devalued women's work (in comparison to that of men) while at the same time normalized women's work in the industry. This type of rhetoric served to, on the one hand, demonstrate acceptance of women's roles in the industry, while at the same time producing a hierarchy of roles and responsibilities that favoured men's work.

Trade press and industry discourses of women in television work during this time were gendered. Women's work in production or technical roles was acknowledged but it was also feminized, for example, by being framed as appealing to female interests. While scholars have since reconsidered and brought value to the work carried out by women in non-technical and administrative work, the discourses produced within the trade press texts I discuss marginalized such work by feminizing it.[61] Betty Goodwin's role as announcer was often referenced in the context of fashion pieces.[62] Similarly, Thelma A. Prescott was heralded as the first female television director (although this was not quite true – it was, rather, a promotional strategy), but this was equally referenced as 'director of women's programmes'.[63] This was one of the conditions under which the women operated. Women, particularly in the post-war years, would be tolerated in those feminine roles or in women's programmes but were to refrain from reaching beyond this. As Robert S. Alley and Irby B. Brown note, 'the same government that had urged women to enter defence work was telling them, by 1945, to go home and do what women are supposed to do – cook, clean, and rear children'.[64]

As the institutional model of television expanded and formalized, women were largely marginalized through a process of exclusion from production and the devaluing of roles that were administrative or supportive (and that many women occupied). Television producers and directors such as Thelma A. Prescott and Helen Sioussat had prominent production roles in the early years of experimental television production but had left by the late 1940s and did not have careers in post-war television. Others, such as Frances Buss and Lela Swift, worked as directors of many early experimental television programmes but contended with an industry that considered them only qualified to direct 'girls' television' after the experimental years, a suggestion which was, as Swift explains it, used to devalue her work in daytime women's television.[65] Indeed, by this time, 'in scheduling terms and in cultural association, daytime was generally denigrated as a feminized discursive space'.[66] In other words, the trade press's enthusiasm for women's work did not translate into industry enthusiasm for women's work. On the one hand, the publicity machine of early television gave the impression that women had access to roles in television production; yet, on the other hand, women experienced barriers to entry based on their sex. It is important, nonetheless, that they contributed as much as they did during these years. While some women were employed in order to fill positions in the absence of men, and others were utilized as promotional gimmicks ('first female director' label for those such as Thelma A. Prescott), their work in early television in the years prior to commercial television undoubtedly shaped what the mass medium would eventually become.

In the formative years of US television, then, broadcasters seemed keen to champion and celebrate their female production staff in the wider press even if this championing did not result in a good deal of actual work for women. The many announcements about, and significant press attention paid to, Thelma A. Prescott are representative of this. Prescott was hired by NBC to direct programmes aimed at a female audience and, as William Boddy suggests, this interest in women's programmes stemmed from concern by 'leaders of the new television industry … [at] integrating television programming into the routines of the housewife's daily chores just as radio had done'.[67] Even before the start of commercial television in 1941, broadcasters anticipated its arrival and aimed to model the formats and genres of television on those already existing in radio. The announcement of Prescott's employment formed part of the overall promotion of television as a medium responsive to women. Like Betty Goodwin, Prescott had experience of media work prior to taking a position with NBC. She was a writer for *Women's Wear Daily* as well the *New York Herald* in Paris and, like Goodwin, had experience in publicity, running her own publicity bureau for hotels.[68] Her experience of writing for women's fashion publications was assumed to make her especially suited to direct programmes for women. Press articles worked to foreground her expertise in women's interest subject matter as well as her professionalism as a television director. Much was made of her status as the 'first female director' even though she might more appropriately have been called *one* of the first directors of television (Frances Buss was also directing during the same period). *Broadcasting*, for example, offered the following introduction:

> Television's first woman program director, Miss Thelma A. Prescott, has been added to the staff of NBC to represent the feminine interest in this new art, it was announced Jan. 23 by Thomas H. Hutchinson, director of television programs. Miss Prescott will produce fashion shows and other programs with appeal primarily to women.[69]

The same magazine referred to her technical work in February 1939 where she is named as aide to the director Warren Wade in a feature on the technical organization of a television show. Prescott and Wade were said to 'study scanning and production technique under all sorts of conditions'.[70]

NBC's internal employee publication, *NBC Transmitter*, made much of her prior experience and education, outlining in detail her years abroad, her work in publishing and other sectors. 'A photographer, journalist, fashion expert and artist, Miss Prescott will handle events and programs of interest to women and will assist in the direction of programs "from the woman's angle".'[71] A number of magazines carried a photograph of Prescott by a broadcast camera, with script in hand and pointing towards off-screen. This served to prove Prescott's ability to direct by showing her go through the motions for the camera. In both *Short Wave Television*[72] and *Radio Craft*,[73] these were accompanied by reference to her status as 'first female director'. This worked to reinforce her expertise and skill. By showcasing her in the act of undertaking her job, the magazines suggested that female professionalism would be accepted and nurtured in the television industry. Given that this was at the very early stages of television and predated the public's first direct access to television at the World's Fair, this promotion of Prescott might have suggested that the public were to understand television as a more equitable industry that included women not just as viewers but as makers of programmes. In the press photograph of her directing television, Prescott was represented in a position of authority, with the photograph suggesting that she was issuing a command to an off-screen person, perhaps to a member of the television crew.

However, Prescott's career in early television never materialized. She directed one programme in 1939 called *Girl About Town*[74] and left NBC shortly thereafter. In an *NBC Transmitter* announcement, Prescott was said to be transferring from the Television Production Division to the NBC Artists Services Department to deliver lectures on working in television production. It stated that:

> Her new work will take Miss Prescott to women's clubs, drama schools, universities, etc., throughout the East. The talks will vary according to the groups' interests, but will usually include various anecdotes as well as more technical information on such subjects as lighting, staging and directing.[75]

One promotional brochure made much of Prescott's extensive experience of television direction and production. While these brochures were, no doubt, intended to persuade aspiring television production staff, advertisers and the public of Prescott's expertise, they also contributed towards a discourse of women's professionalism. Prescott, the brochure noted, had 'written original scripts, adapted stage plays, cast actors and actresses in dramatic roles, planned scenery, selected costumes, supervised rehearsals, determined camera operations and directed countless other details of Television Production routine'.[76] While this perhaps exaggerated Prescott's short time and limited work in television, it also worked to convince the reader that women were capable of production work. Cary O' Dell suggests that her departure was the result of wider layoffs of television personnel undertaken by NBC at this time.[77] However, she did go on to promote television and television work in lectures after her departure from television directing and, as the *NBC Transmitter* announcement declared, these were aimed at female audiences. In particular, she emphasized in interviews that women would be afforded equal access to television jobs since it was an emerging industry.

In a 1940 interview, it was suggested that she thought of television as 'a brand new industry untouched by the tradition of male dominance' and proposed that the interest of advertisers in the television industry would inevitably necessitate female participation in the medium.[78] The article stated that:

> Women will also be let in on costume designing, supervising productions, checking details, acting, dancing, training actors in their parts, and assembling programs, she points out ... And [Prescott] thinks that with men and women both novices in the field – as they are bound to be in anything as new as television – the women will have a real chance to get into important roles.[79]

This feature on Prescott is perhaps indicative of the contradictory position women were faced with in the emerging television industry. On the one hand, the outwardly presented façade of television – evident in the press and promotional material – positioned women as a valuable and necessary part of the budding industry. On the other hand, their participation was contingent upon, and regulated by, gatekeepers who determined the extent of the part they would play. Prescott's image as a female figure representing women's television worked as a marketing ploy for NBC. The company could generate a certain amount of press attention around the 'novelty' of a female director and it could announce to the public its intention to develop women's programmes. NBC invited its female audience to imagine NBC television as television by and for women. Since television was largely unavailable at this time, beyond experimental television in a few urban areas, this need not have been practically demonstrated at this time. By heralding Prescott as the expert in women's programmes, NBC fashioned an image of television programmes as inclusive, accessible and progressive.

Ultimately, however, Prescott directed little television. In her role delivering lectures to women's clubs and societies, she continued to fulfil a marketing function presenting television as an industry and medium that were open to women. The reality, it seems, was that as television became an ever more significant force, women were addressed more as audiences and less as potential employees. Prescott went on in the 1940s to set up a production company, Padula Productions, with her then husband Edward Padula who began working at NBC television in 1938.[80] In 1948, they attempted to get a number of projects off the ground, including *World's Letters* (never developed) and *Girl of the Week*.[81] They succeeded in finding sponsorship for *Girl of the Week* with hosiery and underwear company, Julius Kayser. *Girl of the Week* aired between September and December 1948 on the NBC network. Prescott was listed as director-narrator-writer.[82] The programme profiled fashion models initially and, later, female athletes in five-minute segments that aired at 7.45 pm.[83] Reviews of it praised Prescott's direction but criticized her narration. 'Miss Prescott has done a workmanlike job on the production ... Miss Prescott, however, would do well to delegate the narration; her own voice isn't too well modulated for the purposes.'[84] It may have been an isolated criticism but it reflected age-old gendered language about women's voices. *Girl of the Week* was, according to *Variety*, expensive to produce at $1,000 per episode and was dropped by Julius Kayser after the first run of episodes.[85] In 1952, it was sold in a package of failed TV series and pilots.[86] Prescott continued to write scripts with her husband into the 1950s; however, she didn't return to television or media beyond this point.

Trade discourses of women's work in television, as demonstrated in the case of Prescott, were part of the identity formation of the emerging television industry. These discourses did not materialize independent of Prescott. She shaped discourses of women's work by contributing to articles and features on her own directing work and on women's roles in television. However, Prescott may not have had much control over the trade discourses about her place in the industry since her career was so short-lived. It was not this way for everyone. Others such as CBS broadcaster Helen Sioussat carefully crafted a professional identity and image by offering public lectures and talks on women's roles in broadcasting and by publishing her guide for aspiring broadcasters, *Mikes Don't Bite*.[87] Sioussat's foray into television was much more short-lived than her extensive career in radio broadcasting. Her route to experimental television was perhaps more conventional than Prescott's since Sioussat already held a prominent position in CBS. Yet, while she helped to develop one of the most significant formats – the talk show – that would dominate US television in the post-war years, her first and only television production, *Table Talk with Helen Sioussat*,[88] finished in 1942 after forty-eight episodes. This was less a testament to the quality of Sioussat's programme or her status as a female programme-maker and more the result of CBS's decision to discontinue most broadcasting in 1942.[89] In fact, as CBS executive Roger P. Smith remembered, nobody was particularly convinced that the talk show would make for good television. As he recalled, 'Helen arranged talk shows. Nobody paid any attention to her. Talk? On television? What a misuse of a visual medium! Poor Helen! We nearly blushed when this relic of radio passed among us.'[90] Smith later went on to revise his position and acknowledge that Sioussat was a pioneer in the television talk show format. Nonetheless, when television broadcasting recommenced Sioussat remained in radio and had a prolific career that extended into the 1950s. She started in radio in 1934 and she was assertive in pursuing her first position in radio.[91] She quickly gained more experience and responsibilities in a short number of years, and eventually managed all productions of radio producer Phillips H. Lord.[92] Moving to CBS in 1935, she worked first as assistant to Edward R. Murrow on Talks and, following Murrow's departure, as Director of Talks and Public Affairs. This was a title that Sioussat had to pursue through the company's legal department since there were a number of men at CBS who were resistant to women in senior roles.[93] Equally Sioussat fought for pay parity and she noted that in her early days at CBS she was expected, as a woman, to take on the responsibilities of the role without corresponding pay.[94]

Sioussat significantly expanded the Talks and Public Affairs range of programmes for CBS. Her tenure in CBS radio saw a substantial increase in talks output and she was formidable in managing and gatekeeping the subjects, contributors and discussions on CBS's talks programmes.[95] As a result of her skills in radio talks, Sioussat was well placed to develop a television equivalent for the broadcaster.[96] Sioussat, like Adams in the

BBC, was one of the first people to develop television talks programme, CBS's *Table Talk with Helen Sioussat* from 1941 to 1942 for W2XAB, an experimental station in New York City. During these years, there was little ownership of television sets among the public and CBS considered television programmes more generally, and the *Table Talk* programme more specifically, to be an experiment in what might work if television were to become a popular medium. CBS had, in fact, cautioned its few viewers about the quality of the television programmes and sought to manage their expectations of CBS television.

> We hope you enjoy our programs. The Columbia Broadcasting System, however, is not engaged in the manufacture of television receiving sets and does not want you to consider these broadcasts as inducements to purchase television sets at this time. Because of a number of conditions which are not within our control, we cannot foresee how long this television broadcasting schedule will continue.[97]

Table Talk was developed as an experiment in the talks format rather than a formal effort to attract audiences. Sioussat had been asked to 'work with [the idea] and find out what we could do to make it interesting [for] people to see' in relation to the development of a roundtable talks format for CBS.[98] Equally, the programme was developed as much as part of CBS's effort to fill airtime and to address FCC concerns about CBS's requirement that affiliates carry its sponsored programming.[99]

Table Talk was generally composed of a panel of representative contributors related to a particular public affairs issue, often related to current affairs such as the war. As part of CBS's general ethos, the programme structure foregrounded balanced discussion and debate, with Sioussat acting as moderator of the conversation. Given Sioussat's experience and expertise in radio talks, she was able to secure important public figures as programme guests and to establish a format and structure that enabled this balanced discussion. And, like Adams' BBC television talks in Britain, *Table Talk* was intellectually stimulating and informative. Because of the various impacts of the war on the development of television, and CBS's reduction in television production, *Table Talk* was among the programmes cancelled in 1942.

However, despite the postponement of the development of television more broadly, *Table Talk with Helen Sioussat* helped CBS create foundational practices in format production that it could continue with in later years. Sioussat continued in CBS for many years after *Table Talk* in the field of radio broadcasting as Director of Talks & Public Affairs, and she maintained an interest in women's role in broadcasting throughout her years as a member of American Women in Radio and Television. Following *Table Talk*, Sioussat rarely worked in television again. She was occasionally recruited to participate in CBS television political programmes; however, she did not continue producing television.[100] Indeed, Sioussat's legacy has been as a political journalist more so than a television producer, largely because of her involvement in political reporting.

Sioussat was assertive in establishing her seniority within CBS, which, at that time, had few women in production roles. In addition, Sioussat was careful to represent CBS as a largely fair and equitable organization and to give the impression (to other aspiring broadcasters, or to others in the industry) that her sex had little impact on her experiences as a broadcaster. Her largely positive experiences of CBS were not always mirrored by other women: Shirley Wershba, for example, recalled that she took it as normal that women could not present the news. News production was, for Wershba, primarily the terrain of men.[101] Ruth Ashton found that a shift from CBS radio to television in 1948 meant a drop in status and a limit on her journalistic work.[102] In contrast, and despite the lack of women in CBS executive roles, Sioussat insisted that she didn't encounter gender discrimination from at least two key figures at CBS whom she had close working relationships with: William S. Paley, CBS president, and Edward R. Murrow, who was at that time Director of Talks. She stated that 'I didn't find that [Murrow] felt that men were superior, and he was always very encouraging. And if you didn't do exactly the right thing, he would be patient about it ... he was a good teacher.'[103] In this sense, Sioussat used these allies as a way of demonstrating that, given the chance, women could work alongside men without issue and to the same standard.

She did, however, experience quite a bit of resistance from other senior executives who, she maintained in an interview years later, did not like women in executive positions.[104] Nonetheless, by referencing individual instances of resistance to women or individual executive's disdain for professional women, this contributed to a discourse of an egalitarian organization welcoming of women. Instances of sexism were examples of 'bad apples' rather than systemic.

Perhaps because of her experiences, Sioussat's narrativization of her own career and of women's careers in broadcasting drew upon images of stoicism, restraint and meritocracy. When Sioussat addressed women in public lectures or in her 1943 publication, *Mikes Don't Bite*, she was careful to insist that women take responsibility for their own career progression and learn to negotiate the gendered broadcast workspace.[105] Although this individualized what was a structural problem of gendered work cultures, it nonetheless functioned to assure women that they could work in broadcasting if they worked hard enough. This discourse of meritocracy framed her own career as well-deserved and as representative to other women of what they could achieve if they pursued careers in broadcasting. Sioussat was, in this way, a public advocate of women's access to, and equality within, the media industries and she insisted in *Mikes Don't Bite* that there was space for women in the broadcasting industries.[106] In the book, Sioussat offered advice to women aiming for more senior roles in broadcasting:

> You've got to 'show 'em' girls. Not by affecting low-heeled shoes and masculine garb, by acting self-important or bossy. If you have the stuff of which executives are made, don't be tactless or step on other toes or over the boys ahead ... Be modest and natural; never fluttery, hysterical nor late. Win respect for your abilities rather than demand it. Men can be your most valuable guides, as well as your most powerful foes.[107]

Sioussat was more than aware of the challenges women faced in the broadcast industry and acknowledged that 'as a general rule, girls start as stenographers and end as stenographers'.[108] She also, at times, reinforced those very gender stereotypes that she wished to challenge, as was the case, for example, when she insisted upon having female-only assistants since she thought it humiliating for a man to act as assistant to a woman.[109] However, her advice helped women to manage the male-dominated broadcasting work culture and to prepare them for the inevitable hurdles that would be placed on them by fact of gender.

While Sioussat made some reference to the resistance among senior executive men to female authority, this didn't prevent all women from establishing careers first in experimental and later in commercial television. Frances Buss, who began her career as a temporary assistant with CBS in 1941, was quickly moved to experimental television, starting first in front of the camera and then undertaking a variety of roles behind the camera, ultimately working as a television director. Buss offers an example of one of the few women to make the transition from experimental to commercial television. In an *Archive of American Television* interview, Buss remembers being unimpressed with television when first seeing it at the World's Fair in 1939.[110] Her route into television was largely accidental. While acting as receptionist at CBS, she had the opportunity to contribute to news programmes as she could draw and produce maps for shows (such as during the televised news of the Pearl Harbor attacks). Her background in theatre meant that she was comfortable in front of the camera and, in her early years, she acted as quiz score-keeper and interviewer across a number of programmes under the management of Worthington Miner. When CBS reduced and then ceased its television productions in 1942, Buss was let go and went on to work as a crew assistant with Willard Pictures, making navy training films. Here she gained experience that would enable her to become a director when CBS re-hired her in 1944. From this point onwards, Buss was assigned various director jobs across a wide range of genres, including variety, quiz, sports, children's and women's programmes.

As Buss recalled, CBS (as NBC had done with Prescott) made much of her status as the 'first female television director' and used this to generate publicity for the company (see Figure 7).[111] Like Prescott, then, Buss's status as a female director was exploited for the purposes of optics. It provided CBS with stories to tell about developments in television.

Figure 7 CBS television director Frances Buss at work in a CBS Television studio. Image dated: 27 April 1948 New York, NY, with permission of CBS/ Getty Images.

As a female director in a largely male domain, Buss received a lot of attention. She was featured in a print advertisement for Life Insurance in 1950 in which she was praised as 'another do-it-yourself-American', who worked her way through the ranks to become 'the only woman director-producer on CBS-TV' and refused 'to let the fact that this was "a man's job" discourage her'.[112] This press attention was a double-edged sword of sorts. It served to acknowledge the participation of women in the emerging television industry but, within some individual reviews and features, also implied that any shortcomings were a result of her gender.

While most directors – male and female – were critiqued in the regular reviews in *Billboard* magazine, much reference was made to Buss's lack of skill. For example, a 14 April 1945 review of an evening's viewing in *Billboard* criticized Buss's direction of *Missus Goes a Shopping.*[113]

> The Missus Goes a Shopping, under Frances Buss' directorial hand, was no better and no worse than it has been in the past. Some of the gags were awfully tired and Miss Buss tried to follow people in spots where she shouldn't. But, over-all, it wasn't bad at all.[114]

A 15 June 1946 review, again in *Billboard*, noted that Buss was yet to demonstrate skill:

> At CBS (WCBW), Worthington Miner, top legit director and station boss has stepped out of actual shot calling and as a result there aren't many dramatic segs on the sked. Most of these recently have been handled by Frances Buss, only gal megger [a term for director] at New York stations. Miss Buss can always be expected to do a competent scanning, altho her Casey, Press Photographer (sole recent straight play), from an acting point of view, was less than good. Buss is rated as okay all-around director, but thus far hasn't been given anything that would call for special abilities ... and hasn't shown anything especially creative.[115]

Even in those cases where Buss was praised for her skill, discourses of gender still materialized in the reviews.

For example, a 1944 review of one of CBS's first programmes following its return, *They Were There*,[116] both acknowledged Buss's authority producer and reinforced gender norms by differentiating between good technical work and poor craft work. While the reviewer commended Buss's capable production, it cited the technicians as the real experts. 'There is nothing amateurish about the way WCBW technicians go through their paces ... They start a show on time, are plenty help about the handling of cameras and equipment.' In contrast, the make-up department was said to need improvement in order to make the female actors and performers' faces more telegenic.[117] Buss later stated in an interview that she was sure her gender had an impact on the type of roles she was offered.[118] This was echoed by one of Buss's protégés, director Lela Swift, who recalled that her promotion to director was accompanied by an informal clause. 'The head of the programming department called me in and said, 'They're going to make you a director. ... But I think that you know you'll only be doing cooking shows "cause you're only a girl".'[119]

Buss, nevertheless, was central to many of the innovations in television and continued with her CBS career until the 1950s when she resigned and moved to New Jersey. She was part of the development of colour television, directing the commercials that accompanied the first colour programmes. She was also involved in the development of the Radio and Television Directors Guild in the late 1940s. One of the few women who did not get excluded from television, she was able to continue directing a range of programmes, despite the tendency for women to be restricted to women's programmes.[120] For example, Buss directed one of CBS's early daytime television shows, *Vanity Fair*: a largely female-produced programme that was described as 'a welcome exception to most run-of-the-mode TV programs for women'.[121] Buss was one of a number of female television directors who were able to work in television at a time when it was not yet formalized and institutionalized. Because television was in its infancy and was still on the periphery of the main broadcasting industry, and because the war years resulted in a lack of men available for work, women found less systematic resistance to their employment. Those such as Buss and Swift leveraged their access to the new medium to gain the experience and expertise that would enable them to continue working in the industry after the war. This was undoubtedly a difficult task, since the formalization of television production resulted in barriers of entry and opportunity to women. In other words, when men were once again available for work after the war and when television began to be taken seriously as a medium, gatekeeping practices – which worked to perpetuate gendered patterns of labour – emerged.

This process was nowhere more evident than in the case of station WBKB in Chicago during the period of 1942 to 1948 where, due to the shortage of male staff resulting from the war, women undertook a number of technical and creative roles, including station manager, director, producer, writer, presenter and camera operator. WBKB was by no means an anomaly in respect of its recruitment of women during the war years. In fact, NBC (as well as many of its affiliate stations) had actively recruited women in roles left vacant by men during this period. NBC was, by 1943, already training women in technical, administrative and service work and heavily promoted opportunities for women in its staff publication *NBC Transmitter* where women were celebrated and championed for their undertaking of 'men's work'.[122] In 1944, Margaret Cuthbert, acting as chairperson of the women's activities division of the public service department at NBC, undertook a survey of women's work in NBC stations as well as its affiliates. She found women working in a wide range of administrative, technical and production roles, including '36 directors of women's activities, 14 women program directors, 6 station managers, 11 traffic managers, 21 directors of continuity, 28 women announcers, 12 control operators, 43 women who plan, present and conduct their own programs and many others'.[123] In other words, as was the case in the wider US economy, the broadcasting industry momentarily opened its doors to women in roles previously restricted to men. NBC, for example, evidenced this practice across its broadcasting network and did much to generate a discourse of inclusivity at that time.

In the following discussion of experimental television station WBKB, I account for the complex discourses about the WBKB women's work that emerged from the station's own promotional and marketing initiatives as well as the wider trade press. Initially, at least, the press discourses about women's work were aligned closely with the 'Rosie the riveter' image, whereby the women were admired and their work presented as a novelty. The recruitment drive that resulted in the employment of women to cover men's roles in producing, station management, and technical and engineering work was promoted in the press as a crucial and useful resolution to the manpower shortages that were projected to occur that year. However, reviews of the resulting female-produced WBKB television programmes show occasional intolerance towards the women's perceived inability and lack of experience. Even though the women produced a huge volume of television programmes, reviews were often gendered when it was considered of poor quality. It was, perhaps, no surprise that male expertise was brought to WBKB once again when it was possible following the war. However, the brief history of WBKB's female-led station demonstrates the potential that women had when they had access to television production. While WBKB did not commit to further employment of women, its retention of some of the women coupled with its continued recruitment of women as writers, directors and announcers suggests that the WATTS team convinced the WBKB owners and management of their competencies. The bigger loss, then, was that the employment of women was deprioritized despite their achievements and skill and in order to make way for men.

The station started as W9XBK and operated from 1939 onwards under the ownership of Balaban & Katz, whose main business was cinema theatres. As the war commenced, the station found itself understaffed due to men joining military service. Station manager Bill Eddy also offered its space and staff to the war effort. Yet, determined to maintain broadcasts in order to retain its experimental licence, Balaban & Katz decided to recruit and train women to take on those roles left by male television crew. Balaban & Katz exploited the press interest in the recruitment of women, using them to demonstrate the facilities at the station, to advertise its Radar School, to promote its excellent engineering staff and to advertise its schedule of programmes. *Broadcasting*, for example, announced that:

> The Balaban & Katz television station, W9XBK, Chicago, will operate for the duration [of the war] staffed 1005 in both production and technical departments by women. Six of the station's engineers have been inducted into the Navy en masse as special instructors in the Navy Radar School, which occupies space adjacent to the television station … About No. 1 the station, under the direction of Helen Carson, will be on the air nightly from one to 1½ hours with live talent, dramatic skits, lectures on ceramics, by the Chicago Art Institute, studio wrestling matches, news programs etc.[124]

It posted a job advertisement in local newspapers in 1942 which read: 'WANTED: Telegenic talent girls for technical work in television studio. Mechanical experience unnecessary.'[125] Helen Carson, who had been promoted from secretary to Station Manager, recruited seven women at first (eventually rising to around fifteen) from the hundred women to respond to the ad. These women formed the Women's Auxiliary Television Technical Staff (WATTS), a group which operated at W9XBK, later WBKB, between 1942 and 1945. After the end of the war, some of the women remained, while others continued to work in television and some left the industry.

While training was promised to the new female employees, WATTS member Jean Minetz recalled that most of the training was brief and that, effectively, the women learned by doing.[126] After an initial three-month hiatus on WBKB broadcasting, during which time the women learned the craft of television production, programmes were once again aired. While the women undertook tasks and jobs as necessary, some specialized in particular areas. The women employed during these years included camera operators Rachel Stewart, Esther Rajewski, Marilyn Rosenberg and Jean Schricker; engineer as well as responsible for personnel Marge Durnal; Jean Minetz who worked on audio; director/producers Pauline Bobroy, Gladys Lundberg, Loretto Pagels, Beulah Zachary and Lorraine Larson; and director Fran Harris.[127] Once the women had demonstrated their abilities, they were tasked with developing a full schedule of programmes for afternoon and evening broadcasts. With Carson as manager, the women were encouraged to submit programme ideas and, if it was possible to deliver them within the small budget, they were often produced.[128]

Magazine reviews of WBKB programmes and articles about the station collectively worked to acknowledge the contributions the women were making to television and to note the difficult circumstances under which they were operating (including bad equipment, a small studio space and no budget). For a while, at least, the women were treated as professionals and as innovators in television. Although the reviews were often mixed, they tended to avoid pinning any failures in programme execution on the programme-maker's gender. Instead, like the reviews of other stations' programmes, the reviews outlined the programme subject, genre, the competencies of its execution and identified those who were responsible for the production. A *Variety* review from 7 February 1945, for example, reviewed an hour-long evening production and named the performers and contributors as well as the female directors and camera operators.[129] The review acknowledged the constraints under which television operated and praised the work of Fran Harris in producing the drama *Bright Star Shining*.[130]

> Best thing on this program was 'Bright Star Shining', a half-hour dramatic sketch written and produced by Fran Harris. Based on a good idea, which could have been projected more completely if elaborated upon, the play lost its effectiveness by being patterned to fit the studio's limited facilities. At that, Miss Harris is to be commended for displaying marked craftsmanship in unfolding her story, a simple plot concerning temptations that beset wives of soldiers overseas. She kept her cameras moving fairly smoothly, with no waits, but more thought could have been made on scenic effects and lighting for a better picture. Camerawork was very light and hazy.[131]

Billboard magazine provided a regular review of WBKB's weekly programmes and many of the women are individually named throughout the years 1944 to 1946. While the reviewer, Cy Wagner, seemed to have a particular fondness for presenter Ann Hunter, he was perhaps less biased in reviewing the work of the production team. For example, in a scathing 15 April 1944 review, he noted that:

> This was a day on which the entire B&K staff 'should have stood in bed'. Everything that could possibly go wrong at a television station went wrong … In spite of valiant efforts on the part of the production and engineering staff, the program was hopelessly bad.[132]

Further damning criticism appears in a 7 July 1945 review where Wagner noted that a good programme was let down 'because the station direction staff and camera crew did their best to ruin the show with mistakes'.[133] Wagner went on to outline in specific detail poor lighting, poor timing of slides and the obvious advertising in a cookery programme. However, Wagner also offered praise to individual women for writing, directing and producing good programmes. A 28 July 1945 review praised Carson's direction of drama *X Marks the Spot*,[134] particularly the innovative effects used in the programme delivery.[135] A 28 September 1946 espoused the talent of director Beulah Zachary whose *Music for You*[136] programme demonstrated 'how video can be top entertainment if it is produced and directed with imagination, ingenuity and ability to use all the medium's techniques'.[137] However, by this stage, attitudes in the trade press were shifting. By late 1945, returning men were taking up employment at WBKB and station manager Bill Eddy had returned in 1945 to drive forward WBKB as a commercial operation.[138]

The trade press saw this as an opportunity to professionalize the station and television production and began to frame the station's television output during its early years as rudimentary and the women as mere stand-ins for the returning men. Whereas the women had once been considered saviours of sorts and encouraged to take the helm at the television station, the prospect of the return of a male-dominated television station had reviewers recasting the women as the station's problem and not its solution. One May 1946 report suggested that performers invited to take part in WBKB television productions were angry about the low-quality productions and the disorganized station.[139] One performer was reported to have stated: 'There are too many people, especially women, running around trying to be big shots and treating the rest of us like they're doing us a big favour.'[140] Another was quoted as saying 'We'd like to learn [about television], but we're not ... Something could be made of television in Chicago, but there's too much incompetence from a bunch of girls, most of whom have had no stage background and don't even know the rudiments of direction.'[141] The women, more than anything, were singled out for criticism and were represented as part of the problem that the station needed to solve. By 1946, the women's work was part of a press narrative that represented the station as moving from rags to riches and from amateur to professional television production. Reporter Frank Morgan highlighted the terrible circumstances under which the station barely operated during the war whereby wartime WBKB television had 'been on a hit-or-miss basis, not by choice, but because of circumstances'.[142] Among these circumstances was that 'the station itself was then staged with women camera operators, technicians, directors and execs'. As Morgan tells it, the station was turned around once the man-in-charge, Bill Eddy, had returned. By this time, 'broadcast technique is being improved, as are some of the regular programs, and the staff is being strengthened'. In effect, Eddy was 'cleaning house' and solving those problems of inexperienced, female staff by incrementally replacing them with men.

Douglas Gomery states that the women 'did a fine job under trying circumstances, but when veterans returned starting in the fall of 1945, Balaban fired the women and filled the position with returning male soldiers'.[143] However, it was not the case that all WATTS employees were fired and at least a few of the women were retained as writers, director and producers. It seems that it was those women in technical roles who were largely replaced by men. What we can gather from the WBKB weekly programme reports in early 1946 is that there was a concerted effort to replace technical and craft roles undertaken by WATTS with men. In addition, there were concerns about having female producers and directors manage male crew. A report from 28 January to 1 February 1946 from second-in-command at the station, Reinald Werrenrath, to Bill Eddy stated that:

> By hiring men for the actual operating studio work, we should be able to achieve greater efficiency in most cases. However, if we plan to have our present staff of directors be directly in charge of these men when doing studio programs, we will run into trouble, as no more than three of our present directors are, in my estimation, capable of running a studio of men.[144]

This would appear to be a concern with female directors and female authority over male technical and production staff, since the majority of the WBKB directors were, at this time, female.

Further on in the report, it was evident that the women at WBKB attempted to retain some of their authority in this space:

An alternate or transitional solution suggested by Marge [Durnal] to meet our immediate problem would be to put one of the senior girls in charge of control room operation and riding the script for an entire evening.[145]

This was perhaps a suggestion to move women out of the live production environment and into the control room where they could assist and offer technical support from the control room, with the remainder of the production team in the television studio. This negotiation was, in all probability, strategic. The returning men had the technical qualifications that justified their employment in technical roles. By 9 February 1946, Werrenrath proposed a formal delineation of organizational structure, roles and titles for studio production. This was 'proposed to facilitate our transition into large scale commercial broadcasting'.[146] Since many of the women weren't officially titled any specific role other than WATTS and since they undertook a variety of roles, this was something of an inhibition and made their positions somewhat redundant.

This campaign of recruiting more men and of instituting formal staff titles and roles had begun in 1945 and continued throughout 1946. This coincided with the departure of some of the WATTS women from the station. For example, in September 1945, three of them left the organization to work in other television and theatre companies. These included programme supervisors Helen Carson, engineer Marge Durnal and public relations staffer Ann Drobena. At the time of their departure, two men were hired to supervise educational programmes.[147] This was followed by the recruitment of former radar officer and 'video vet' Dave Crandell as producer in January 1946.[148] In February, WBKB hired five male engineers and technicians following their return from the navy. A *Billboard* article also notes that such hires would not impact current staff. 'Additional expansion of WBKB's staff is expected to take place soon. In the near future, [Bill] Eddy plans to hire new personnel to take over camera assignments and other studio jobs. Present studio staff will not be discharged but merely be enlarged.'[149]

According to O'Dell, neither Minetz nor Harris knew how and why WATTS ended or, in fact, whether it was officially disbanded at all. Since some of the women stayed on, it seems that they weren't fired from WBKB and some retained employment at the station, particularly as directors and producers.[150] In addition, up until at least 1949 Esther Rajewski and Rachel Stewart were still working on camera as evidenced in WBKB promotional material and *Variety* and *Billboard* reviews from that period.[151] However, by the late 1940s reference to women in technical roles began to fade.[152] And in trade magazines, there was little reference to them beyond the 1940s. Stewart and Rajewski received a brief acknowledgement in the promotional booklet 'WBKB – Looking Ahead with Television' (see Figure 8) where it stated 'the unique distinction

Figure 8 'WBKB Looking Ahead with Television', 1948, p. 26. The photo is accompanied by a caption that uses the 'first women to ... ' as evidence of the station's pioneering role in Chicago television. Pictured are Esther Rajewski and Rachel Stewart.

of being the only feminine television camera operators in the country belongs to Esther Rojewski [*sic*] and Rachel Stewart, who came to WBKB during the war and stayed'.[153]

The most significant impact on women's technical work at WBKB in the post-war period, as elsewhere, was that there was a need and a desire to reemploy men back into technical work at the stations. In addition, and as demonstrated in at least one WBKB internal document, mixed-sex working environments were thought to pose a problem, and female authority in technical and production work wasn't immediately accepted by or acceptable to male staff. The women who worked in technical or production roles at WBKB into later years were those who were recruited during the WATTS period. Very few, if any, were recruited after this time. Once the war was over, the enthusiasm for and celebration of women in television, particularly in technical work, faded, and there was little press attention on the scale of that of the wartime period. Without this 'normalization' or encouragement of women in technical roles, it was perhaps less of an incentive for women to pursue such work. Women could no longer 'man' television. At least to the same extent. It is perhaps telling that some of the women who learned their craft at WBKB went on to be leaders in the field of television. Fran Harris Tuchmann had a successful career in advertising, being the first woman in the advertising industry to lead a TV department. Lee Phillip, who was employed by WBKB in the early 1950s, went on to develop legendary soap operas *The Young and The Restless* (1973– ongoing) and *The Bold and the Beautiful* (1987–ongoing). This suggests that, given the opportunity to work in television, women could become trailblazers in the industry. Had more women been able to retain and sustain careers in the post-war period, the industry may have looked very different.

The case of the WATTS women at WBKB offers one brief example of some of the opportunities available in the early years of television as well as some of the barriers to entry that emerged during the post-war period. The very fact that women engaged in the production of television during its formative years indicates not only that women were interested, capable and innovative, but also that their exclusion was part of a, perhaps uncoordinated but certainly encultured, notion that women's place, while not necessarily in the home, was certainly not in those fields prescribed as masculine and including technical operations, production and management. Thus, the institutionalization of US television was accompanied by a reframing of women's roles in relation to television. This institutionalization of television was in place by 1948 when the major networks of NBC, CBS, ABC and DuMont developed programming and scheduling practices that were supported by and facilitated the commercial model of television.[154] The foregrounding of female television talent and skill that had marked the experimental years of television was replaced with an emphasis on women as consumers of television, and women largely disappeared from the discourse of television production. As Alley and Brown note, 'beginning with the earliest days of television network decision making, women were rarely in the board rooms. And the product communicated to the American public reflected only a tiny role for women in writing, directing and producing.'[155] Those women who sustained or built careers in television in the post-war years were, for the most part, exceptional. Women were largely excluded from the sphere of production and instead were invited to participate in television as consumers of it. In this way, women were still included in television culture, albeit outside the masculine domains of technology, industry, business and production. In the next chapter, I trace the origins of this turn towards the woman as consumer and the subsequent emergence of the female audience in the broadcast industry.

Populations, consumers and audiences

The concept of the female audience emerged very early on in the development of the broadcasting industry. Market research reports, listener and viewer studies, and audience research were carried out by a range of companies, agencies and institutions from the 1920s onwards. The discourses about audiences and their behaviours, as well as the tools for carrying out research on audiences, were borrowed from academic, market and consumer research that arose in the late nineteenth and early twentieth centuries.[1] Indeed, if the purpose of television audience research and measurement was to understand what audiences engaged with and how, it is useful to examine how audience research originated, particularly in relation to the production of the category of woman in both commercial and academic research.[2] The emergence and expansion of consumer capitalism and the concomitant growth of the fields of market and consumer research resulted in the disciplining and shaping of consumers. First, consumers were quantified and then classified according to their perceived social groups and interests. Researchers then sold the results of their studies to other brands, companies or advertisers. Gender was an important category in this emerging consumer economy. Women in particular were addressed as central participants in the consumer economy, either as naïve victims who were subjected to the will of marketing experts or as pragmatic practitioners of rational consumption. In the case of the former, scholarship on the female consumer figured her as representative of the masses, a derogatory term that signified the passivity of individuals within a sophisticated system of production, distribution and sales.[3] In the latter case, academic literature addressed how the rise of consumer culture enabled women to have agency both in the domestic sphere and beyond.[4]

These polarized perspectives on the female consumer are representative of the complex and changing discourses of womanhood and femininity in consumer culture. Such gender discourses were produced (or resisted) by many actors and institutions including the media, marketers, social researchers and, of course, women themselves. In the early twentieth-century, academic scholars, psychologists and social science researchers had begun to study female markets and consumers.[5] This was suggestive of the wider efforts to classify women's role in the economy and to produce them as a social category that was meaningful to advertisers, businesses, marketing companies and media organizations.[6] We can ask what kind of 'woman' emerged from this research and measurement. What motivated the focus on white, middle-class women in market research? What were the processes by which knowledge was generated about women as a category? What forms of knowledge were generated in early market research and how was this enacted and put to use in the early broadcasting industry? This chapter examines the rise of market and consumer research and the discourses of the female consumer that were produced by such research. It identifies how the female consumer was transformed into the broadcast industry's female audience.

In her assessment of the relationship between gender and consumption, Victoria de Grazia points out 'that there was nothing natural or inevitable about the development of modern consumption practices' which produced a 'dichotomized relationship' between men as producers and earners and women as domesticated consumers.[7] She notes that when domestic technologies and devices entered the home, a new form of household emerged, one that had the female as its central figure. De Grazia suggests that there are two feminist cases to be made about women and consumption: one that positions women as victims of mass consumption and another that 'argues that mass consumption liberates women by freeing them from the constraints of domesticity'.[8] Indeed, the new consumer culture both limited and created opportunities for women. Consumerism offered women some agency, authority and access to the public and commercial sphere. At the same time, however, this was contingent upon women's identification with a gender identity constructed by the consumer and media industries. This consumerist identity was the product of market and consumer research and, as de Grazia suggests, not an inevitable result of social and technological changes. Rather, this 'dichotomized relationship' was a product of concerted effort within the consumer and media industries to gender consumer practices. This was legitimized and normalized in market and consumer research, which seemed to offer scientific objectivity, methodological rigour and extensive analysis of women's consumption. Feminist scholarship on market and consumer research has been concerned with the ways that scientific objectivity as well as the production of knowledge have been gendered masculine.[9] This is not to suggest that women do not contribute to knowledge production (for example, many market and consumer researchers were women), but that 'since historically men have largely been the sole proprietors of knowledge ... and since objectivity is socially perceived as a valued masculine trait, it is easy to see how the perception of an objective, unbiased and ungendered perspective could be created and perpetuated'.[10] For example, the common use of the term 'housewife' in early consumer and market research was indicative of research bias which favoured groups of women who were married, worked in the home and were not engaged in paid employment outside the home. This effectively worked to reject, or treat as anomalous, the many women who did not conform to the category of housewife.

Feminist approaches to consumer research emphasize the way in which knowledge of female consumers is produced through 'historically, socially, and politically shaped discourses' that normalize certain consumerist ideas and ideals of men and women. Consequently, 'the language of consumer research includes several terms for female consumers such as "house-wives," "work-wives," "just-a-job wives," "career women," and "super-women"' where there are 'no corresponding set of terms ... for male consumers'.[11] Women in early market and consumer research were defined largely by their marital status and their domestic roles as mothers and carers, whereas men were treated as a default group. Early market and consumer research evidenced such practices and focused intensely on producing ever more research literature that profiled female consumers for the consumer industry, advertisers and media organizations (much of which was contradictory, narrow in focus and centred mainly on white middle-class women). This research resulted in a gendered opposition between masculine marketer and feminine consumer. Feminist scholarship has interrogated the way that consumer research may be 'partially reliant on gendered signifiers for their meaning because the taken-for-granted linguistic opposition between marketer and consumer is related to our culture's binary opposition of male versus female and masculine versus feminine'.[12] Early market and consumer research used gender in such a way: as an interpretive framework through which to gain knowledge about consumers. Early market and consumer research 'othered' women by subjecting them to regimes of knowledge that transformed them into female consumers.

We can turn to the market and consumer research of the early twentieth century to identify the ways that the female consumer was discursively produced as a social category. Market and consumer research's practices of gendering respondents and participants influenced the broadcasting industry and the various forms of audience research that emerged in this industry. During the early twentieth century, those engaged in the production and sale of consumer products became more interested in identifying potential consumers. In response, advertising agencies such as J. Walter Thompson (JWT) and N.W. Ayer & Son started to carry out research on the public and its consumption practices, such as those consumers of cars, newspapers or household products.[13] Therefore, much of what was researched and measured in consumer's practices, lifestyles and choices was tied closely to consumption and economic activity. Important were those factors in a consumer's life that could be quantified and made meaningful within the context of consumerism, such as economic status and social class, purchasing power, gender, age, race and ethnicity but not those things that could not be quantified and measured in market and consumer research.

Charles McGovern traces the commencement of this form of research to 1910 when advertisers in the United States began to undertake the formal study of consumers. 'Led especially by J. Walter Thompson, agencies theorized, located, surveyed and interpreted consumers.'[14] If agencies had more knowledge of the consumer, so the logic went, they would have the power to persuade and to steer the consumer towards certain goods and products. In Britain, as in the United States, organizations such as JWT embraced market and consumer research and had by the 1920s employed experts in psychology and business to carry out such research for companies and brands. These experts encouraged JWT to consider the class and social composition of consumers in their marketing practices. According to Stefan Schwarzkopf, 'Because classes constituted ordered divisions in a society whose members shared similar values, interests and behaviours, it was vital for the agency to be able to separate its audience according to socioeconomic classifications.'[15] As McGovern notes, 'advertisers continually sought to classify people by income levels, occupations, regions and tastes and to correlate such observations with markets.'[16] Gender was quite an important category for advertisers, especially as it intersected with race and ethnicity and social class as well as marital and employment status. In an era in which women were becoming more visible in public life, the image of the female consumer produced by the advertising and consumer industries stressed and amplified gender differences but treated women as a cohesive group.[17] McGovern notes that between the late nineteenth century and the Second World War, many in the advertising industry believed that women – middle-class, white women – held significant purchasing power, particularly in terms of purchases for the domestic sphere.[18] While women were valued as purchasers of consumer products, their status as socially and biologically inferior remained.

In addition, the image and profile of the female consumer produced by market, consumer and audience research worked to negate or neglect demographic differences among and between women and among the wider population. Those women who represented a valuable market for advertisers and brands came to be understood as the typical consumer. According to Roland Marchand, consumer citizenship was granted by the advertising and consumer industries to those representing the ideal social class that consisted of those with enough purchasing power to make them attractive to the market and consumer industries.[19] Those who typically did not meet this socio-economic threshold – including minorities, for example – did not and were not counted and, therefore, represented. Marchand points to the paradoxical situation in which the term 'mass audience' often represented quite a narrow population of white, middle-class consumers.[20] This is evident in the market, consumer and audience studies that effectively excluded those from working-class or minority backgrounds who did not meet the sufficient economic threshold that turned them into consumer citizens. Marcel Rosa-Salas demonstrates how early twentieth-century market researchers in the United States used all-white survey samples that were taken as 'nationally representative' and, therefore, effectively excluded African American and minorities from a consumerist mode of address.[21] For Rosa-Salas, 'early mass marketing efforts were a form of racially targeted advertising to White Americans of economic means'.[22] Marchand suggests that this 'averaged' consumer citizen was not only white, but was almost exclusively imagined as a 'she'.[23] The consumer produced by market, consumer and audience research and then addressed by advertisers, brands and in the wider media was 'defined and limited by race, class, and ethnicity but promoted now as "average"'.[24] The white, middle-class, married homemaker that came to mean the average female consumer, in fact, represented only a small segment of the female market. Nonetheless, she became a significant participant in the consumer industries.

Once this female consumer was produced, a paradoxical situation emerged where women were perceived to have some form of 'masculine' rationality and agency through their purchasing power but engaged in what was ultimately defined as 'feminine' consumption.

This manifested in advertising campaigns that addressed women as emotional and in need of advice or, alternatively, as pragmatic and knowledgeable purchasers. As Simone Weil Davis notes, 'during the early decades of its ascendance, the industry vacillated between soft sell-atmospheric, "mood" advertisements and hard sell-fact-driven "reason why" ads; these were, by and large, the "revolutions" and struggles that gripped advertisers'.[25] In other words, advertisers alternated between representing their consumers as emotional and feminine or rational and masculine. While women were sometimes addressed as rational, the female consumer was largely perceived to be emotional. 'Advertising envisioned women according to business interests, recasting women's work in male terms while promulgating an ethos of consumption as a naturally female trait.'[26] These perceived natural traits followed common sense assumptions about female tastes, interests and characteristics, and these assumptions, in turn, shaped how women were understood as consumers and addressed by advertisers. McGovern suggests that 'Most male advertisers believed that women were less capable of "reason" than men, and therefore, successful advertising to women would have to be based on suggestion and sentimental depictions, images that aroused feeling without a conscious chain of reasoning'.[27] Advertisers and market researchers considered women as more open to suggestion and, consequently, an appropriate target for advertisements and promotion of consumer products.

In addition, the female consumer was imagined through the lens of the products pitched to her and the commodified environment that she operated in (the household, the family, beauty and fashion). For example, by the 1930s US radio soap operas were accompanied by sponsorship from brands such as Mickelberry Products and Brown Beauty Baked Beans,[28] both of which were sponsors of daytime soaps aimed at a female audience. The female consumer was, therefore, imagined as domestic – motivated by household work and household management. For McGovern, advertisers grappled with this dual approach to women because they were both 'beholden to millions of women' but also patronizing of them. 'If advertisers privately derided women's housework, ad men quickly acknowledged her purchasing power.'[29] For McGovern:

> These professionals thus were compelled to profess loyalty and dedication to women while feeling at every turn superior to them. Such dependence on the habits, interests, budgets, and attention of supposedly emotive and illogical women fostered resentment ... That same alienation colored advertisers' relationships and perceptions of the mass public of consumers.[30]

While advertising agencies may have felt beholden to the public, the introduction of market and consumer research techniques promised to enable those agencies as well as product manufacturers, retailers and media organizations to access the right consumer markets through the generation of extensive knowledge about such markets.

Market and consumer research was produced and published widely in the early decades of the twentieth century. Articles and books such as Harlow Gale's *On the Psychology of Advertising* (1900)[31] and Daniel Starch's *Advertising: Its Principles, Practice, and Technique* (1914)[32] were used to aid advertising agencies and advertisers in gaining an understanding not only of how advertising worked but also the consumer market.[33] Early pioneers of market research included Boston's Curtis Publishing which, from 1911, had a division of commercial research under the leadership of Charles Coolidge Parlin.[34] Curtis was the publisher of newspapers, magazines and journals such as the *Saturday Evening Post, Ladies' Home Journal* and the *Country Gentleman*. Its research division carried out survey research on consumers and markets in order to more successfully sell advertising space in its publications.[35] Parlin standardized the use of interviews, surveys and questionnaires across sample populations and, during his first few years, compiled a number of reports that assessed and demystified consumer habits, attitudes and tastes. He gathered data on consumer expenditure, tastes and tendencies and aimed to provide a scientific and reliable overview of the consumer market. Gender became a key concern of such research. For example, in 1914 a survey on automobiles claimed that women go on to become key decision-makers in car purchases, something that may have seemed unlikely at the time of the survey.[36] Market research firm Emerson B. Knight had an advertisement in advertising trade journal *Printers' Ink* that claimed that the 'proper study of mankind is man … but the proper study of markets is women'.[37] Research divisions and enterprises like those in Curtis Publishing or market research company Business Bourse began to undertake this 'proper study' by using quantitative and qualitative consumer research methods and by using statistical data that was drawn from public sources (e.g. tax registers) or through subscriptions. However, while rigorous research methods might have been used to evidence total consumer populations and wider trends, the specific characteristics of the female consumer were often poorly evidenced, based as much on anecdote, researcher bias and observation than on qualitative data.

One such example was Parlin's report *The Merchandising of Textiles* published by the National Wholesale Dry Goods Association in 1914.[38] The report assessed consumer trends in shopping trade areas with a particular emphasis on the differences between men's and women's purchasing habits and decision-making processes. Parlin spent some time accounting for the methods by which the report had calculated figures for potential trade in a given shopping area or department store. However, claims about gendered consumption were evidenced not in 'hard data' but much more anecdotally, for example, from third parties such as department store sales managers. And where 'hard figures' were used to represent volume of trade and estimates of shoppers, no such figures or quantitative data were used to represent the claims made about the female consumer. While the report did attempt to highlight women's savviness as consumers as well as the rational process by which they made purchasing decisions, it nonetheless produced less an account of the habits of actual women shoppers and more a conceptual profile of an imagined female consumer. As Patricia Johnston argues, much of the early literature that made supposedly scientific claims about gender differences among consumers was absent of any evidence or justification.[39] She suggests that market and consumer researchers produced a rationalized yet largely unsubstantiated profile of a female consumer who conformed to the traditional and socially sanctioned role that woman had in society.[40] The use of methods found in the social sciences and psychology gave the impression of academic legitimacy and rigorous investigation. Yet, while many of the key market and consumer research books and reports did much to emphasize good research practice, many of the claims made about women's consumption were based on speculation and intuition.

This is evident in perhaps one of the most seminal industry texts on the female consumer: home economist-turned market researcher Christine Frederick's popular and ground-breaking *Selling Mrs. Consumer* of 1929.[41] This book, as Marcy Darnovsky suggests, 'combined the ideological and economic justifications for consumerism into a formula that doubled as a manifesto of American superiority'.[42] *Selling Mrs. Consumer* foregrounded the role and participation of women in the consumer economy and made them a central figure in this economy. The focus on the female consumer was intended to demystify women for advertisers but it also had the effect of producing a traditionalist and conservative image of her. Published by Business Bourse, the market research company headed by Frederick's husband J. George Frederick, the book made claims about the female consumer that were suggestive of comprehensive and original research and investigation. Yet these claims were largely based on a few narrow studies that had been conducted by a third party – in this case, a study[43] on a small sample of New York families by market researcher Harry Hollingworth.[44] Although Frederick was eager to emphasize women's expertise and authority in the home, her *Selling Mrs. Consumer* was directed not at women themselves but at the largely male-dominated consumer industry. The profile of the female consumer she offered was that of a homemaker and, while she stressed the power, intelligence and agency of the female consumer she also represented women as domestic, maternal, white and middle-class.

Throughout the book, Frederick stressed the necessity of valid and systematic research and she took advertisers to task for poor research of female consumers. Frederick used US Bureau for Standards figures on household spending to support her psychological profile of the female consumer which was largely restricted to household management (in other words, largely homemakers or housewives).[45] She criticized advertising and marketing men for having ideas about female consumers that were not supported by hard data.[46] She advocated for market research to be undertaken by specialized organizations rather than universities and attacked the naivety of some manufacturers who simply guessed women's interests and tastes when questionnaires and surveys would have provided more concrete data.[47] She was also cautious to acknowledge the limitations of statistical reasoning and estimation. "'Average' individuals or families do not exist, of course. They are statistical abstractions ... Still, it is a necessity in selling to find common denominators and visualize a composite picture of the mass, or at least certain particular levels of the mass.'[48] Elsewhere, Frederick noted that the use of class categorization for female consumers may not have accurately reflected the diversity that existed within these groups. 'There is plenty of diversity within each class or level [of consumer], because of the great geographic spread of these United States and the recent foreign origins of so many of our people.'[49] In other words, Frederick advised on cautious use of market research data.

However, Frederick went on to classify women according to a range of criteria without providing any evidence or data to support her claims. For example, she classified women by age groups that were reflective of the supposed consumption trends within this group. Between the ages of sixteen and twenty-two, women were said to be narcissistic and driven by vanity, pleasure and personal adornment.[50] Women between twenty-two and twenty-eight were in the 'romantic home building period', and from twenty-eight to thirty-eight women were engaged in 'alert Home-making, Cooking and Housekeeping' making them 'most approachable and suggestible, alert and open-minded'.[51] From age thirty-eight, women were claimed to be concerned with 'luxury and comfort, travel, new housebuilding'.[52] In offering a comprehensive list of traits and characteristics of the female consumer, Frederick perpetuated stereotypes about, and also produced distinctions between, the genders. This was largely unproven by Frederick, despite the occasional reference to other studies, reports and surveys. Frederick produced seemingly coherent and stable categories of women that all suggested a middle-class housewife and homemaker. This particular representation of the female consumer was perpetuated elsewhere, where, in studies of gendered consumption, women were largely referred to as housewives (e.g. in William J. Reilly's *Marketing Investigations* of 1929[53]). The housewife, in this way, came to be understood as the archetypal consumer who could be explained and accounted for by market and consumer research and for advertisers.

This female consumer of market and consumer research became the female audience of broadcasting, particularly in the United States where audience research was driven by the commercial model of the early radio industry. In the United States, and later Britain, the earliest forms of formalized audience research were less concerned with listener profiles and more concerned with the number of sets in use. During the early 1920s, there was little attention to individual listeners and their habits and tastes. Researchers instead used the household or family as the basic unit of measurement. Because the industry was in its infancy during the early 1920s, it was more important to establish that there was an audience in the first instance. While radio programme-makers had at this time solicited correspondence from listeners in order to gauge general programme interest, more systematic audience research was required. Throughout the 1920s and 1930s, there was a growth in social and audience research and measurement accompanied by the development of techniques such as surveying, polling and interviewing, particularly in the United States. These tools were used to measure, interpret, understand and shape the public. As Sarah Igo suggests, the research undertaken on the public was motivated by a number of concerns. 'Carefully collected data could be used to assess economic conditions, tap efficiently into public opinion, guide national policies, and perceive social reality more clearly.'[54] The development of such research methods was connected to the rapid growth of populations, of consumerism and of media usage. Broadcasting emerged at the same time as these methods were being developed and, indeed, many of the methods – from ratings systems to surveying – were later used as a means to measure the ever-increasing audience for radio broadcast.[55]

In the United States, there was a special need to measure the audience since radio was established as a commercial, advertising-supported business and, thus, broadcasters wanted to both gain knowledge of their own listenership and, from there, turn this listenership into a commodity to justify prices for advertising space on the station schedules. While US broadcasters in the late 1920s received quite a large volume of correspondence from listeners, this did not provide the same level of data that was already produced on consumers of print media.[56] The radio audience was frequently referred to in broadcast magazines and trade journals as the vast 'invisible audience' suggesting the lack of knowledge about who was listening.[57] As early as 1923 the *Wireless Age* had issued a questionnaire to stations to determine the volume of letters received from listeners in order to attempt an estimate of the total audience.[58] As advertisers demanded more accurate listener data, methods such as survey and consumer interviewing were adopted from the fields of market and consumer research.[59] As early as 1928 broadcasters, such as NBC, commissioned one-off surveys to determine how many people were listening, where they were listening from, and what their general socio-economic background was. NBC's 1928 survey carried out by market researcher Daniel Starch attempted to discover the socio-economic status of the network's listeners.[60] However, the Starch survey did not have a wider impact at NBC or in the advertising industry.

Instead, as radio broadcasting and radio audiences expanded in the United States, more regular measurement of audiences was required. This was facilitated by a number of companies, including the Cooperative Analysis of Broadcasting (CAB) formed in 1929 by Archibald Crossley and Clark-Hooper in 1935 (which developed the 'Hooperatings' measurement system). Broadcasting audience research was also undertaken by other market researchers, social scientists and academics, and pollsters. George Gallup, for example, undertook research of print, broadcast and film audiences using his public opinion polling methods. Paul Lazarsfeld worked with a number of public organizations and media companies to produce market research on radio audiences. As demand for audience research grew between the late 1920s and early 1940s, interest in and research on more detailed audience composition came to the fore resulting in more profiling and categorizing of the total audience. The household unit was replaced as an object of study by individual family members. The mother then emerged as an important figure for audience researchers and advertisers.

The use of the household as a single unit of measurement initially enabled audience researchers to be strategic in how they used and represented the data. In addition, conceptualizing audiences in terms of household income meant that they could be selective in terms of the samples used, thus making the resulting research more attractive to broadcasters and advertisers.[61] Audience researchers often collected listener data by phoning radio listeners and asking about their listening. Because those households with telephone subscriptions were more likely to be of higher socio-economic status, this produced a narrow sample that gave a particular representation of the household: one that was consumerist, with the head of household in paid employment and the woman at home. In other words, the sample consisted of families with a specific middle-class gendered division of labour. While the ratings reports and surveys produced by Crossley's CAB and Clark-Hooper's Hooperatings represented their data in terms of households, they nonetheless worked to profile this household in terms attractive to advertisers and broadcasters who were interested in understanding the potential consumer market for radio.

The practice of measuring individual family members was, thus, a strategic one. This formative audience categorization was useful for broadcasters who needed to convince advertisers that they had a high quantity and quality of potential consumers among their listeners.[62] For example, a 1927 survey 'Radio as an Advertising Medium' used the family household as the basic unit of measurement while asking questions about what time each household member listened at and what member of the family operated the radio set. The family members were identified as 'father', 'mother', 'young people', 'children' and 'other' in the first set of questions and in a later question as 'husband', 'wife', 'young people' and 'children'. A question on programme preference referenced both family membership and socio-economic status.[63] In other words, there were efforts to differentiate the listening habits of individual family members. Ratings organizations, broadcasters and advertisers produced a concept of the audience that was suggested to represent the entire listening public. However, this audience was reflective of a particular segment of the population that was attractive to advertisers. By focusing on the ideal consumer market, ratings organizations painted a picture of a household that was made up of keen consumers and organized by gender.[64] In turn, the broadcast industry produced schedules and programmes that were thought would lure the audiences that advertisers wanted. If, in high earning households, housewives listened during the day, schedules reflected this. Thus, programmes appealing to women were scheduled, not when all women were watching, but when women with consumer power were watching.

Given that the housewife (assumed by advertisers and researchers to have consumer power) was a key demographic for advertisers, many surveys were undertaken in an attempt to shine light on how she engaged with radio. However, these surveys and reports demonstrated a preoccupation with housewives' daytime listening even if those same surveys found that women listened more during the evening. In addition, they made some assumptions about women's programme preferences that were not always supported by data or were contradicted in different surveys. Kellogg and Walters' 1932 survey 'How to Reach Housewives Most Effectively' emphasized the need for advertisers to give serious consideration to housewives' daytime listening habits since they

> make up a large and important part of the radio public. The program
> sponsor should realize that the housewife in a majority of cases is the
> member of the family who has the most influence upon the family

purchases and is the one who spends the greatest amount of time in the home. She is, therefore, the member of the family most easily reached by radio broadcasts.[65]

The researchers thus borrowed from the logic of market and consumer research which held that the housewife was the primary purchaser in the household. Based on 900 questionnaires that were issued to housewives, the study concluded that 'evening hours are the best time to make an appeal to housewives' since 95 per cent of them listened at this time.[66] Kellogg and Walters claimed that housewives also listened during the 2.00 pm to 4.00 pm slot and 9.00 am to 10.00 am and proposed that the reasons for these listening patterns were that women were engaged with various housekeeping tasks at other times. Kellogg and Walters ranked listening time preferences as 'evening first, mornings second and afternoons third'.[67] Despite recognizing the popularity of the evening slot, Kellogg and Walters offered detailed suggestions about the most suitable daytime slot for sponsors and advertisers, noting the barriers to listening, and the likely attentiveness of, housewives at various points in the daytime. This preoccupation with the housewife's daytime listening habits was perpetuated or referenced directly in a number of other studies.

These studies collectively worked to normalize the association between the housewife and daytime listening. In 1935, Pauline Arnold produced the study 'Sales Begin When Programs Begin' for NBC, drawing on data generated from the listening diaries of 3,042 housewives.[68] This study was carried out 'to demonstrate the significant level of daytime listening by housewives, and that listening often accompanied or preceded activities in which advertisers' products were used'.[69] Frederick Lumley's 1934 book *Measurement in Radio* cited Kellogg and Walters' claim that housewives indicated appreciation for instructive programmes such as recipes, child care and education.[70] It also repeated Kellogg and Walters' claim that 'the housewife is the one to whom radio advertising should be directed'.[71] Warren Dygert's 1939 study for NBC, *Radio as an Advertising Medium*, suggested to advertisers that programmes that appealed directly to women – including 'service programs on cooking, domestic economics, home decoration, health and beauty' – would all work in gaining the attention of the housewife.[72] In Herman Hettinger's 1933 book *A Decade of Radio Advertising*, reference was once again made to the daytime listening of women.[73] It noted that 'with regard to the time of the day, the afternoon hours constitute an almost unlimited field of development for the broadcast advertiser who is interested in the female listening audience'.[74] Hettinger went on to suggest that, since housewives will have completed their daily chores by afternoon, they might prefer to relax with programmes such as 'a restful semi-classical concert, popular music done in recital rather than dance style, or an interesting dramatic sketch with sufficient romantic appeal'.[75] Hettinger maintained that, by 1932, women were no longer interested in women's instructional programmes and were becoming tired of 'broadcasts of recipes, household helps, child-care information, suggestions regarding interior decoration, beauty hints and similar features'. He suggested that women were 'demanding more in the way of entertainment'.[76] However, this view was contradicted not only in Kellogg and Walters' study but in psychological research on radio audiences.

Hadley Cantril and Gordon Allport's 1935 book *The Psychology of Radio* made a number of claims about gendered listening, one of which was that women's tastes were framed by their domestic roles.[77] While it noted some general preferences among both sexes' listening habits, it suggested that 'there are marked differences of taste depending upon the age and sex of the listener. Many programmes that appeal to men do not please their wives and daughters'.[78] Among those programmes said to be enjoyed by women were soap operas, symphonies 'and, of course, fashion reports and recipes'.[79] Men, on the other hand, preferred more serious and professional programmes with their preferences listed as 'sports of all kinds, talks on national policies, detective stories, talks on engineering, physics or chemistry, and business news'.[80] However, the narrative account of men and women's preferences was rather different than the data provided in the charts and supplied in the aggregation of questionnaire data.[81] While the narrative suggested that women favoured women's genres and programmes, this was not entirely supported by the ranked list of women's genre and programme preferences, where fashion and recipe programmes, for example, appeared lower down the list.[82] In fact, the ranked list indicated that women preferred football programmes over fashion reports and had an equal preference for baseball and recipe programmes.[83] In other words, the tables provided in the study suggested that men and women's tastes were not oppositional. However, the conclusions drawn suggested that women's tastes were highly gendered, and different to men's tastes.

The discourse of gendered radio listening prevailed and came to be one of the fundamental 'truths' about radio audiences. Audience research defined the female listener as a housewife who listened mainly during the day and to programmes that were aligned with dominant ideologies of femininity. Despite some research evidencing either large variations in women's listening habits or similarities in listening for both men and women, the 'common sense' ideas about the female listener were perpetuated in further studies throughout the 1930s and 1940s, such that the daytime hours, being 'women's hours', were dominated by 'women's programmes'. The discourse of the female audience was also reinforced in social sciences research that was increasingly attending to the role of radio in society and the meaning it had for the listening public. More generally, where market research was concerned with mapping how and when the public listened, the social sciences were concerned with *why* the public listened. However, this distinction was not absolute and, indeed, many of those engaged in communications research for universities and public bodies also worked in the communications or broadcasting industries.

One of communication research's leading figures, Paul Lazarsfeld, had worked with the Office of Radio Research (a social research division based in Princeton University and later Columbia University) in the 1930s. This was funded by the Rockefeller Foundation for the purposes of carrying out audience research for commercial clients.[84] Frank Stanton, who went on to work with Lazarsfeld in the Bureau of Applied Research at Columbia University, worked in audience research with CBS. The Bureau of Applied Research went on to publish a number of widely referenced studies on radio in the United States throughout the 1940s. These studies were framed as sociological and academic yet had many of the hallmarks of market research and, in many cases, were driven by the agendas and perspectives of commercial clients.[85] It is perhaps no surprise, then, that the findings of the studies mirrored many of the findings of commercial audience research. And, like earlier studies such as *The Psychology of Radio*, the claims made about women's listening often sat in opposition to the data that supposedly evidenced such claims.

Lazarsfeld's 1940s study *Radio and the Printed Page*, for example, used the language of social sciences to justify many assertions and assumptions about women's habits, preferences and, in particular, their tastes in regard to radio.[86] Lazarsfeld demonstrated a preoccupation with cultural taste, and this informed much of his analysis of class and gender. Lazarsfeld, as an example, delineated between serious and light listeners and serious and light programmes and found correlation between the 'decrease of serious listening' and 'the decrease of cultural level'.[87] Regarding gender, Lazarsfeld understood women in traditional roles of housewife and mother and did little to map the wider female audiences for radio. This was apparent in his persistent association of women with housework and mothering, even while he made a case for more complex and meaningful understanding of audience behaviour elsewhere. Lazarsfeld also noted an intersection between class and gender, claiming that women's preference for magazines or radio was evidence of their social-economic status. For Lazarsfeld, reading was superior to radio and those women who read homemaking information were, in turn, more qualified in and capable of their housekeeping than those who listened to radio. 'Women who have a more genuine interest in housekeeping – those who regard it more as a matter of craftsmanship than of routine work – are more likely to prefer magazines rather than radio under otherwise similar conditions.'[88] While Lazarsfeld went some way towards outlining how both reading and listening occur in higher numbers among women, his polarization of men and women, as well as urban and rural listeners, demonstrated a taste and cultural hierarchy that positioned women as less serious and less sophisticated consumers of news media.

This somewhat contradictory construction of the female audience was evident throughout a number of publications by Lazarsfeld and others in subsequent years. In Lazarsfeld and Stanton's publication, *Radio Research, 1942–1943*, gender differentiation was central to the production and interpretation of audiences.[89] Three chapters were dedicated to the daytime serial and offered a variety of methods of researching and interpreting audiences. Each attempted to make sense of the gratifications afforded by daytime serials and to identify the motivations for listening either through sociological analysis of female listeners or through content analysis of serial programmes. Researchers Herta Herzog, Rudolph Arnheim and Helen J. Kaufman all exhibited, to varying degrees, a condescending attitude towards the female audience of daytime serials, although Herzog and Kaufman treated this audience with more seriousness, perhaps, than Arnheim. For Kaufman, an understanding of the daytime serial audience needed to be supplemented with contextual information regarding the production and broadcast of such serials.[90] For Herzog, assessment of the audience needed to be situated within extensive qualitative research of the actual listening public. For Arnheim, who was more concerned with the content and form of daytime serials, the female audience was a passive group at risk of suggestion by the manipulative drama of the serial.[91] While each attributed different levels of agency to the female listener, they all took for granted that gender differentiation was an inevitable part of audience research. While Herzog and, to an extent, Kaufman questioned the extent to which gender alone contributed to listening habits and preferences, none considered why gender was the primary means of categorizing listeners.

Herzog was perhaps somewhat more nuanced in her approach to listeners of daytime serials. She rejected the cultural elitism of her colleagues and peers who considered the daytime serial a trivial and low taste form. Drawing from four separate studies on daytime serial listeners, Herzog used a three-pronged approach to daytime listeners: studying the content of serials, comparing listeners and non-listeners, and engaging in qualitative audience research on listeners. Herzog offered a number of hypotheses or common assumptions about the daytime serial audience and subjected these to testing and scrutiny. She also questioned some of the methods of obtaining data, such as the nature of the survey questions, and sought to situate some of the survey conclusions within a wider contextual field. More generally, Herzog sought to demonstrate that the female audience was diverse, with different interests, tastes and concerns and that definitions of any female audience would always necessitate more research and investigation than the quantitative research methods used to date. Among her conclusions was that daytime serial listeners tend to have less formal education than non-listeners. Herzog speculated that this was 'because these serials provide the more naïve individual with a much-desired, though vicarious, contact with human affaires which the more sophisticated person obtains at first hand through her wider range of experience'. Moreover, the 'serials which abound in arrays of stereotyped characters and situations are less likely to satisfy those who have more discriminating perspectives'.[92]

Like Lazarsfeld, then, Herzog demonstrated a form of cultural elitism, whereby lower-class women had less taste than their upper-class counterparts. However, Herzog did point to the limitations of some of the surveys she used and noted that the questions asked of respondents were narrow and did not capture the intricacies of women's personality traits as well as their interests. In other words, Herzog recognized that there were far more variables that determined women's listening than the time of day or the genres they listened to. Overall, Herzog recognized that the classification of 'woman' was far too simplistic and needed to be supplemented with further contextual detail. That many women listened to daytime serial meant little in-and-of itself. Women listened with varying levels of interest and enthusiasm. They used daytime serials for different reasons and made sense of them in ways that had as much to do with their social roles, social participation, education and location as their general status as women. While Herzog maintained the use of gender differentiation in audience research, she nonetheless did more to question the commonly held assumptions about the female audience.

Thus, while *Radio Research, 1942–1943* attributed a level of seriousness to the study of radio genres more popular amongst female listeners, it nonetheless demonstrated a general acceptance of fundamental gender differences in radio listening. It then perpetuated such differences by persisting in the use of male and female categories even when differences in listening among men and women were non-existent or miniscule. Women in turn became synonymous with the daytime schedule since more women were perceived to be listening at that time. This foregrounding of women in surveys of daytime listening became so pervasive that the daytime period came to be associated almost exclusively with the female audience (with the exception of children's listening). This occurred despite the fact that women listened in higher numbers in the evening time. However, since the daytime audience had a higher audience of women than of men, research on the female audience was less concerned with evening time listening and more with daytime listening where women could be isolated as a group. For example, Lazarsfeld's 1948 report *The People Look at Radio Again* included tables that outlined programme preferences for daytime and evening listening (see Figure 9). However, where the table for evening programmes included listening figures for total audience, the daytime listening table had viewing figures categorized by gender.[93]

By 1948 Lazarsfeld was so confident in the use of gender categorization in audience research that he claimed that 'sex difference' was 'the outstanding characteristic of the radio audience'.[94] What Lazarsfeld failed to note was that, just as consumer and market research had done in previous decades, such audience research emphasized female difference much more than male difference. What Lazarsfeld articulated in his comment was more a will to discursively produce a gendered audience than a will to understand the complexities of the listening public.

As this chapter has demonstrated, we can turn to the rise of marketing and market research, consumer research and audience research to identify the ways such gender categories were deployed and gender discourses produced in the consumer and later broadcasting industries. Indeed, by the time when radio audience research was undertaken there

TABLE 13

THE CONSTANCY OF PROGRAM PREFERENCES*
(1947 compared with 1945)

	Daytime Preferences				Evening Preferences	
	MEN		WOMEN		TOTAL	
	1945	*1947*	*1945*	*1947*	*1945*	*1947*
News broadcasts.....	65%	61%	76%	71%	76%	74%
Comedy programs....	*	*	*	*	54	59
Popular and dance music............	15	23	35	39	42	49
Talks or discussions about public issues.	22	19	21	22	40	44
Classical music.......	12	11	23	20	32	30
Religious broadcasts..	19	22	35	41	20	21
Serial dramas........	7	6	37	33	*	*
Talks on farming.....	13	16	12	13	*	*
Homemaking programs	5	5	44	48	*	*
Livestock and grain reports............	14	17	6	10	*	*

* Figures do not add to 100% because more than one answer was permitted each respondent. The starred program types are not considered because of the infrequency with which they are heard at the designated times.

Figure 9 Programme preferences for daytime and evening time in *The People Look at Radio Again*, 1948, p. 21.

was a wide body of marketing literature that had produced 'common sense' ideas about the female consumer. This shaped how radio audience research was undertaken. The female audience was, early on, produced as a commodity within the radio broadcasting and advertising market. Whereas the family had been first understood as the basic unit of measurement in audience research, the shift to gender categorization was motivated by the demands of the broadcasting and advertising industry. Both were concerned with identifying a sector of the listening public that was of value to the consumer industry. Using the categories of gender allowed broadcasters to shape their programming according to perceived interests of female and male listeners. Programme-makers in turn could lure this audience by producing content they imagined would appeal to women or men. This was especially the case with programmes addressed to the female audience since women were both imagined as a key consumer market and since the female audience could be traced to specific time slots in the radio schedule. This is evident in the production of a discourse of the daytime housewife radio listener. This listener was the subject of much study, debate and criticism. She was thought to be susceptible to manipulation by broadcasters and advertisers. This may have been good news for advertisers but it resulted in some harsh criticism of women's radio consumption and the demonization of women's tastes. Women were thought to be interested only in trivial and lowly programmes such as musicals, serials and game shows.

Regardless of whether any of these assumptions were evidenced in audience research, they formed the dominant image of the female audience in the radio industry. This concept of the female audience would influence television audience research in the United States and Britain where television audience researchers drew on the same methods, practices and assumptions as those carried out in consumer and radio audience research. In the following chapter, I examine how the female audience emerged gradually in 1940s and 1950s television audience research literature. I emphasize the way that this was driven by the desire of the television industry to secure valuable advertising revenue. In effect, television broadcasters wished to capitalize on daytime television by convincing advertisers that these daytime hours drew a valuable female consumer audience. In order to make their case to advertisers, audience researchers had, by the 1950s, produced the concept of the daytime female television audience. By the time the television industry had professionalized, women's role in the industry was determined to be that of audience and not producer. The next chapter examines the emergence of television audience research and identifies the processes through which the television industry produced the female audience.

The US female television audience

Where other previous public mass media such as theatre, vaudeville and cinema had invited women out of the home and into public space, the experiential space of broadcasting was predominantly domestic. In addition, in the context of public entertainment spaces, women were invited into mixed-sex spaces and shared theatrical spaces, auditoriums and similar venues with men. In other words, women formed part of the audience of public entertainment. Although expectations of gendered behaviour still governed women's use of such spaces, it is important to note the extent to which these enabled women to stake some claim on public spaces through the act of being part of an audience. In contrast, broadcasting was experienced within the domestic space and came to address women in their domestic roles in that space. Broadcasting had the dual effect, then, of shifting the experience of media from the public to the private and, as I suggest throughout this chapter, of developing schedules and programmes that spoke to women in terms of their perceived domesticity. The female audience, perceived as accessible to broadcasters and advertisers precisely because television was a domestic medium, became a significant concern in the television industry. And although broadcasting may have brought the public into the domestic space of the home it nonetheless re-territorialized public and private in terms of gender.[1] This chapter details how gendered audiences became a growing interest in commercial audience research between the late 1940s and 1950s. It argues that audience research reports and studies began to emphasize and, at times, manufacture gender differences in viewing tastes and behaviours in order to sell gendered audiences to advertisers.

To demonstrate this, I examine the discourses of the female audience that were evident in three particular case studies: the surveys carried out and reports issued by television market and audience research organizations and departments of the late 1940s and 1950s, including Advertest, Videotown and NBC's Research and Planning Department. These are selected as representatives of studies that were promoted and circulated in the wider trade and industry press and which were used internally by broadcasters. There were a great many audience studies – both commercial and academic – but not many were widely cited, referenced or used by broadcasters and advertisers at the time. Using these case studies, I demonstrate how, during the late 1940s, there was little reference to the gender of the audience in these studies and reports. Instead, studies of the family as a unit or simply 'viewers' sought to gain an understanding of more general television consumption habits in the home in order to show the impact of television on US domestic and social life, on the consumption of entertainment and on the consumption of products and goods. I argue that during the early to mid-1950s, increased reference to gender occurred primarily in relation to the figure of the housewife or 'lady of the house'. Although the housewife, according to the audience research reports themselves, watched television mostly in the evening (and more than other members of the family), the housewife was of little interest in terms of her evening viewing. Instead the housewife was more interesting to audience researchers as a daytime viewer where, as it happened, she watched significantly less television than she did in the evening. However, broadcasters and audience researchers were motivated, for a number of reasons, to produce a female daytime audience, at least within audience research literature.

By the 1950s, daytime television offered advertising and sponsorship space that was cheaper and more accessible to advertisers than that of the evening time. In addition, broadcasters learned from market and consumer research of the importance of the female consumer and, thus, wished to target her specifically and directly in television advertising and sponsorship. Therefore, since housewives were figured as the main residents of the home during the day, or at least the ones that advertisers sought out, and since the housewife was thought to make many purchase decisions in the home, daytime television for the housewife seemed a logical next step for television. The result was a concerted effort by audience researchers, broadcasters and advertisers to construct, shape and address a female audience that was domestic, white, middle-class and consumerist. This, in turn, had the effect of gender segregating television programmes and television schedules with daytime television for women eventually becoming associated with 'low taste' and trivial programme content in comparison to the evening schedule's association with prime-time or prestige programmes. Prime-time hours, in turn, were the domain of the man of the house, for whom television viewing was a much more serious affair.[2] This chapter examines the role that audience research played in this.

Broadcast schedules were formed around the housewife's supposed domestic tasks. Daytime programmes aimed to address her particular needs and interests, specifically those needs that could be met by consumer products. Television audience research of daytime viewers concentrated mainly on women and, in particular, the housewife. In addition, the television industry put women to work as viewers, since this interest in housewives was driven largely by the potential to sell the housewife audience to advertisers. In the research undertaken by Advertest, Videotown and the NBC Research and Planning Department, most attention was paid to the housewife, with some occasional reference to children and even less to men, or single, retired or working women. Since the housewife was assumed to be the main viewer of daytime television, and since advertisers imagined her as an avid consumer, she became the focus of attention in audience research and in the resulting television schedule that became the norm from the early 1950s onwards. This was a consequence of a number of assumptions about the housewife. Industry and advertisers firstly imaged her as having a key role in purchase decision-making in the family. Broadcasters could, therefore, convince advertisers that a daytime audience of housewives, albeit smaller than the evening audience, was nonetheless valuable. Alongside this, housewives were claimed to form the primary audience for daytime television, since much of their work was carried out in the home and, therefore, in proximity to the television set. If this was the case, regardless of the low quantity of viewers, advertisers could be attracted by the high quality of them. Broadcasters and audience research companies such as those discussed in this chapter set about proving this quality to potential and existing advertisers by carrying out numerous studies on the composition of the television audience, with more and more focus on the daytime television audience of housewives during the 1950s.

However, as the following examination of commercial television audience research reveals, a daytime TV audience composed largely of housewives was manufactured rather than a given. In many cases, studies were carried out with the preformed notion that the housewife was the key viewer. This meant that oftentimes only housewives – primarily white, middle-class ones – were surveyed in studies of daytime television. Studies tended to mask or obscure the low viewing figures for daytime television among housewives, but gave the impression that they watched far more than other members of the family, which was not always the case. In other words, the audience research reported and, consequently, produced rather than reflected gender difference and contributed towards the gender segmentation of the television audience.[3] The female audience produced within television audience research was reflective not of the total female viewing public but of the consumerist ideal attractive to advertisers and broadcasters. The surveys and reports under discussion did much of the work of discursively producing this audience. US commercial television, then, developed in response to a highly selective and imaginary idea of the consumerist, white, middle-class housewife who worked in the home. This was the sum of what female viewers were in the eyes of the early US television industry.

Television audience research emerged out of a particular US broadcasting industry ecology that materialized in the post-war period. Although radio audience research and measurement were undertaken during the 1930s and 1940s, the television industry that developed from the mid-1940s onwards necessitated its own medium-specific audience research, particularly given that television was commercially funded and that the audience was a key commodity exchanged between broadcasters and advertisers. Given the high costs of television production and the high costs for advertisers in accessing airtime, it was important that broadcasters establish large enough networks, a wide enough body of affiliate stations and advertising revenue sufficient to support the pursuit of a television broadcasting industry. The post-war period saw the networks resume television broadcasting after they had temporarily suspended most of their experimental broadcasts in the early 1940s. By 1945, there were now a number of old and new broadcasting networks looking to television as a potentially lucrative new business. NBC, CBS and DuMont were joined by the newly formed American Broadcasting Company (ABC).[4]

Although the networks owned and operated stations of their own, they also sought out other local stations that were interested in expanding from radio into television and that could benefit from network-produced shows. The networks could provide a good proportion of programmes to local stations that might not have had enough content otherwise. Networks looked to affiliate stations to air their programmes and local stations to benefitted from the networks programmes as it meant they didn't need to produce all programmes themselves. Local stations would broadcast a combination of network programmes and locally produced programmes such as news and sports.[5] Networks would often contract their affiliate stations to broadcast the network's sponsored programmes at specific, 'network optional time[s]' in the morning, afternoon and evening. Stations, in turn, would be financially compensated for broadcasting network programmes.[6] The benefit to networks was that they could gain the audience reach that enabled them to attract higher advertising revenue. CBS and NBC were able to develop television networks and to sign up affiliate stations more quickly since they had a track record in radio and established reputations with advertisers.[7] Development of the networks was, however, slow due to technological limitations: without the cables necessary to link networks across the United States, it wasn't possible for some stations to receive network programming and therefore they did not sign up with networks. Audiences for network programmes were, as a result, limited to certain areas that could receive the programmes. Since the funding model of US commercial television depended upon the sale of audiences to advertisers, it was crucial that broadcasters secure the commitment of both: audiences were to commit attention to screens and advertisers and advertising agencies would commit sponsorship and advertising to broadcasters.

Programme sponsorship had been the dominant model of advertiser funding in the commercial broadcasting sector from the 1930s to the 1950s (in the late 1950s, the quiz show rigging scandal effectively brought an end to this model for prime-time, at least[8]). Programme sponsorship enabled advertisers to exercise a good deal of control over programme content as well as schedules.[9] By the late 1940s there were efforts by the networks to shift away from the sponsorship model of commercial television. Under the helm of Sylvester 'Pat' Weaver, for example, NBC-TV began to favour network-made programmes rather than advertising agency-made programmes.[10] However, this was a difficult process since sponsorship was the dominant business model of radio which influenced television. Programmes were also expensive to produce.[11] Since television set ownership remained low until the 1950s, investment in broadcasting and programmes would have to wait for a financial return.[12] Both Weaver at NBC and CBS's owner William S. Paley sought to 'upend broadcasting's traditional relationship to sponsors' by keeping the sponsors' control over programmes and schedules in check.[13] The proposal was to move to a magazine-style advertising system whereby shorter advertising segments would replace one main sponsor, thus limiting the authority of any one sponsor. This proposal was something that advertisers could tolerate so long as the audiences they desired could be reached by the network-produced programme. By the late 1950s, programme sponsorship had declined to around one-third of television programmes, although it remained for longer in daytime television.[14] Once single sponsorship began to decline, 'advertisers enjoyed greater mobility and flexibility in reaching audiences. They were free to follow targeted audiences to the programs that audiences preferred by buying single minutes of time in those programs.'[15] Spot advertising allowed advertisers to focus on individual product rather than general brand advertisements and to expand across the television schedule. It was within this changing industry dynamic that television audience research was being carried out and which was used by advertising agencies and the networks to evidence continued interest in programmes by audiences.

Broadcasters, advertising agencies and advertisers relied on audience research and measurement to evidence audiences for programmes aired and to provide knowledge about audience interests, tastes and behaviours. Audience measurement of the post-war period initially developed in the radio industry with similar ratings services later offered to television broadcasters. Some measurement services such as Hooperatings expanded from radio into television and, in addition, new measurement services including Trendex and A.C. Nielsen emerged in the post-war years. The information offered by measurement companies to advertisers and broadcasters relating to the numbers of viewers for a given programme – the ratings – drove much of the activity of television production and scheduling.[16] Each of these broadcast ratings companies, from the Cooperative Analysis of Broadcasting (CAB) to the Pulse surveys, Hooperatings and Nielsen ratings, aimed to provide the most accurate information on what was viewed, when and, eventually, by whom.[17] Nielsen, for example, claimed to have the most accurate sample of listeners and viewers, although its audiometer device that recorded sets-in-use could not account for who was watching at any given time.[18] The American Research Bureau (ARB, later Arbitron), which was formed by an employee of NBC, James Seiler in 1949, used household viewing diaries produced weekly by survey participants. During the early 1950s, these audience measurement services were more concerned with reporting viewing figures of the audience or programme popularity than with reporting who the audience was. While ARB and other commercial ratings services, such as Videodex, gathered data on audience composition, this was of secondary concern to ratings. Other audience research services operated more like market and consumer research in that these services provided data on audience demographics and composition and related this closely to viewing patterns.

Alongside audience measurement services of the late 1940s and 1950s such as Nielsen, CAB and Hooperatings, other forms of audience research were carried out that more closely resembled consumer and social research. As such, they tended to include more qualitative data on consumers interests and tastes as well as demographic data. Advertising agencies such as J. Walter Thompson formed internal research departments that engaged in a variety of studies of the composition, tastes, incomes and spending habits of television audiences.[19] Market research companies produced regular reports and bulletins on the television audience and that were aimed at a readership of advertising executives.[20] Other studies focused less on programme ratings and more on station or area/local audiences for television. Area studies were carried out using a method of interviewing and diary-keeping developed by Forest L. Whan and were often sponsored by stations. These studies tended to include more reporting on the relationship between audience composition and viewing practices. For example, the 1955 Topeka Television Area Audience Survey was sponsored by local Kansas station WIBW-TV. Data on gendered viewing practices and preferences was gathered and the study ranked the television genre preferences of male and female viewers.[21] Likewise, the CBS-sponsored Iowa Radio-Television Audience Survey of 1955 measured genre preferences by sex but not by age or socio-economic background.[22] An early report for New York state station WRGB and titled 'Some Preferences of Television Audiences' classified audiences in terms of adults and children and focused on programme preferences more than viewing habits or patterns.[23] It ranked programme genres according to preferences and offered little indication of programme choice according to audience classification. The 1950 NBC-commissioned *Hofstra Study* carried out by Thomas E. Coffin at Hofstra College intended to assess viewers' responses to television advertisements.[24] NBC used the study in their attempt to demonstrate to advertisers the effectiveness of television and advertised the results of the study widely.[25] Titled *The Hofstra Study: A Measure of TV Sales Effectiveness*, it referenced the gender of viewers in terms of their responses to television advertisements but did not make much of this in the resulting data interpretation or findings.

At the same time as this research was being undertaken, further commercially sponsored and academic studies were undertaken to gain much more contextual information on the audience. These studies were less concerned with what audiences watched and more with questions of how and why people watch as well as who these people were. These studies also demonstrate increasing concern with the female viewer, more particularly the housewife, to the extent that many of the studies only include housewives as participants.[26] From the late 1940s, then, studies such as Videotown and *Hofstra* began to focus on the contexts and contours of audiences. By the late 1950s when the Videotown studies ceased, the housewife and daytime television were key concerns of many audience research reports. Advertising agencies and advertisers were seeking out more information on the daytime female audience and encouraging research companies to provide such data.[27] Market research companies were providing such reports in the form of studies of advertising effectiveness, women's taste in programmes, women's interest in daytime television and women's consumption habits. Throughout the remainder of this chapter, I assess audience research case studies that are demonstrative of the ways that audience research companies through their reports and studies responded to the needs of broadcasters and advertising agencies for a daytime female audience. Throughout these 1950s studies, then, a discourse of the daytime female viewer emerges: one that is distinctly consumerist and that fulfils the needs of the commercial US television industry.

For Marsha Cassidy, television's expansion to the daytime hours was governed by an interest in the female audience. As Cassidy notes, 'from the first light of television, daytime programmers set their sights on women. Across the nation, producers at the local, regional, and national levels devised a curious assortment of programs calculated to attract the female spectator.'[28] In her book *What Women Watched: Daytime Television in the 1950s*, Cassidy traces the emergence of US daytime programmes for women to, firstly, local television programming and, later, national network programming from the late 1940s. Elana Levine, in her study of US daytime soap operas, notes that advertisers, television producers and even women were initially resistant to daytime soaps in particular, since they threatened to interfere with household work.[29] For Cassidy and Levine, between the 1940s and 1950s, daytime television became synonymous with female viewership; and as US networks began to expand from prime-time to daytime programming, women became increasingly central to the address of daytime television. According to Cassidy, local stations were among the first to broadcast daytime programmes specifically aimed at an audience of women, many of which concentrated on homemaking.[30] Indeed, networks such as DuMont worked hard to convince advertisers that daytime television was a worthy investment. A 1949 DuMont advertisement in *Sponsor* titled 'for daytime television see DuMont', for example, encouraged advertisers to spend on daytime programme advertisements specifically because they had a captive market of housewives. 'If you want to reach the housewife', read the DuMont advertisement, 'daytime television must occupy an increasingly important place in your plans.'[31] For Spigel, national networks were more risk-averse than DuMont in expanding outside of the evening schedule, only making the move in the early 1950s with the increased demand for television advertising space.[32]

The opening up of the daytime schedule did not immediately result in an influx of advertising revenue and it took a few years to make the daytime schedule viable and profitable. Advertisers had to be assured that women – specifically housewives – had the inclination to take time out of their day to watch television.[33] Radio was considered to be much more accommodating since it required less direct concentration and engagement from housewives as they carried out domestic work. Television networks needed to convince advertisers that they had an audience of housewives in the daytime and, consequently, produced programmes that were specifically geared towards women, or more especially the housewife. Since radio broadcasting had already established its presence in the daily routine of the housewife, this may have seemed like an ideal model on which to base television. And, according to Spigel, during the early 1950s 'the major networks were also intent upon designing programs to suit the content and organization of the housewife's day'.[34] Genres that had worked in radio were carefully adapted to television. Soap operas, for example, were developed to emotionally appeal to women and to offer a form of 'therapy' and escape from their daily routine.[35] This genre, according to Levine, 'came to be understood in industry discourse as a means of helping the potentially unstable woman in the home cope with her life'.[36] The industry was able to come to this conclusion by way of various forms of social, market and psychological research. Levine traces the post-war trend in social and, particularly, psychological research that was used to provide a rationalization for public, consumer and audience behaviour. She notes how such research was used in the television industry to evidence an interest in and, indeed, a psychological need for entertaining daytime television for women.[37] In this way, the case was made for daytime television.

Between 1947 and 1951, then, networks and advertisers started to shift towards daytime television and, from 1951 onwards, it was established through institutional audience research that daytime television could be successful and lucrative for advertisers and sponsors since it was demonstrated as having a steady audience of housewives. Therefore, during the years of television's ascendency in US homes – from the late 1940s to the mid-1950s – the institutional concept of the female audience became increasingly important and central to the operational logic of the television industry. As television audience research began to define and shape the viewing public into audiences, the female audience, more often described in terms of the housewife, emerged as a meaningful commodity in the institutional marketplace of television. By Autumn 1951, 'more than half of the TV stations in America' were reported by CBS to be moving to daytime broadcasting.[38] According to CBS the 'most significant theme ... [was] the complete acceptance of the daytime TV by the nation's housewives. [The CBS report's] facts and figures controvert sharply the charge, still heard in some circles, that early-hour video won't get viewers.'[39] The report went on to note that women's daytime television viewing doubled in one year. Audience research did the work of identifying this daytime audience of housewives.

This audience of housewives was not guaranteed simply by providing a daytime schedule. Since housewives were engaged in labour and were, as those in the television and advertising industry were concerned about, already occupied with various domestic tasks and chores, the networks had to work hard to address them. Audience researchers then had to account for them. In other words, the daytime television audience of housewives existed mostly as a discursive phenomenon produced in the address of 'women's programmes' to housewives and in audience research studies' insistence on referring to the daytime audience primarily as housewives. In fact, in the early years of daytime television (up until the mid-1950s), the audience was either non-existent or very small since most potential viewers were otherwise occupied. Many women and men were engaged in work outside the home and older children were at school. Equally included among those few who watched daytime television alongside women were many young children as well as some men. While the daytime audience became synonymous with the housewife, it was, in many respects, a fairly diverse audience. As surveys such as those carried out by Videotown (1948 to 1958) indicated, television viewership remained low even for women who may be in the home. However, if networks wanted to expand their schedules in order to accommodate more advertising, the daytime schedule was the place where this could occur. It was in the networks' interest to address the daytime schedule of programmes to the audience it imagined would be available, even if it was a small one. Thus, the networks began developing programmes that addressed the female audience,[40] including domestic science, magazine-style and shopping programmes and, by the late 1950s, soap operas such as *Search for Tomorrow*[41] and *As the World Turns*.[42] The networks addressed a certain type of audience (one with consumer power) and they produced and commissioned reports on this audience that worked to prove to existing and potential advertisers that the networks had captured this audience. Despite the fact that women consistently watched television in significantly higher numbers, for far more time in the evening and more than any other member of the family at this time of day, networks characterized the female audience as a daytime-viewing housewife. This is evident in the way that women were referred to in the audience research carried out in a number of late 1940s to late-1950s surveys such as the Videotown, Advertest and NBC's own Research and Planning Department surveys. Between the late 1940s and the late 1950s, audience research reports evidenced a shift from the study of the family unit to the study of the woman and housewife, particularly in the context of the daytime viewer profile. Women became increasingly referred to in the studies and surveys as 'wives', 'mothers' and 'ladies of the house', and were increasingly discussed as a daytime audience.

The Videotown, Advertest and later NBC Research and Planning Department studies are indicative of the discourse of the housewife daytime viewer that emerged throughout the late 1940s and 1950s. The Videotown annual studies (1948 to 1958) were produced by advertising agency Cunningham & Walsh and promoted widely in trade journals and magazines. Cunningham & Walsh's clients were sponsors of CBS and NBC programmes.[43] Advertest Research carried out market and consumer research which was sometimes commissioned by and often featured in advertising trade magazines such as *Sponsor*. NBC's own Research and Planning Department carried out audience measurement and commissioned research by other market research companies on NBC's audience composition and advertising effectiveness. In each of the three studies discussed, housewives were the main participants. The 1951 Videotown study was the first to provide data on daytime viewing and this was collected from housewives only regardless of who may have been watching.[44] The 1949 Advertest report 'Study on Daytime Viewing', like surveys carried out by CAB and Pulse, did not overtly state that women only were interviewed; however, all participants were female and reference to participants used the pronoun 'she'.[45] Likewise, the later NBC-commissioned and W.R. Simmons-produced study *Television's Daytime Profile: Habits and Characteristics of the Audience* included women only.[46] In other words, when studies reported on evening or genre programmes, the reports were largely gender-neutral in discussions of the audience (aside, perhaps from some reference to demographic composition of the sample). As more and more reports on daytime television materialized through the mid-to-late 1950s, the concentration on housewives was far more explicit. This was especially evident when reviewing the growing discourse of gendered viewing in individual studies across years.

The Videotown survey demonstrates the incremental attention paid to the housewife in audience research as daytime television became an institutional concern. Between 1948 and 1950, the housewife was barely mentioned in audience research reports. By 1951 (the year that networks turned to daytime television), reference to her increased. Yet the language used, as well as the expression and tone of the surveys, indicated some initial dismay at her limited viewing of television. This was replaced with more optimism as the viewing figures for housewives grew marginally through the years. The ten-year Videotown study commenced in 1948.[47] The town's identity was anonymized at first but later revealed to be Brunswick, New Jersey. The annual studies, eleven of which were carried out, were concerned with the viewing and purchasing habits of people with and without television.[48] The studies were more generally concerned with the changes in television ownership over time, the differences in ownership within socio-economic groups, and the impact of television ownership on the lives and lifestyles of those in the area. The first few years of the surveys contained very little reference to demographics beyond social class and the size of the family. However, by 1950 onwards this changed as the study introduced specific data on the gender and age of both the set-owning and non-set-owning families. Initially the studies had asked about evening time viewing, since that was what broadcasters and advertisers were primarily concerned with, but by 1951 data on daytime viewing was also provided as advertisers were encouraged to turn their attentions to daytime television. Therefore, while housewives watched consistently more than husbands in the evenings as well as throughout the day, discussion of housewives' viewing habits was largely concentrated in sections of the studies concerned with daytime viewing.

The 1948 to 1949 and the 1950 studies were largely focused on set-ownership and socio-economic background, with some reference to audience composition.[49] Reference was made to husbands' and wives' evening viewing habits and the extent to which they viewed or not. It was only in 1951 that daytime viewing was reported on, but only in terms of wives' viewing practices.[50] This study, like the previous year's study, again reported on viewing habits as related to gender and age. When participants were asked about who was viewing in the evening, the study found that the percentage of both wives and daughters over eighteen years old watching evening time television was higher than that of husbands and sons over eighteen. Similarly, wives and daughters watched more hours of television on weekday evenings than husbands and sons.[51] In a section on TV and Social Life, the study noted that daytime television viewing remained low in comparison to radio listening whereas evening time viewing had surpassed radio listening.[52] Daytime television viewing and radio listening were noted only in relation to wives who, it seemed, were the only noteworthy audience for daytime radio and television. They were the assumed audience for daytime television. 'Daytime TV has not supplanted radio in Videotown. In the morning less than 2% of the wives view TV; in the afternoons about 10%.'[53] This concern with the daytime radio and television consumption of women was, in all likelihood, driven by a question about the feasibility of daytime advertising. By 1951, it seemed, radio was still the best medium through which to reach the housewife during the day.

The 1952 study again compared the daytime radio and television consumption of women.[54] The report presented women's continued preference of radio over television as a problem to be solved by providing better daytime television (see Figure 10).

TV VIEWING AND RADIO LISTENING

	WATCH TV		LISTEN TO RADIO	
	1951	*1952*	*1951*	*1952*
TV OWNERS				
Wives				
Morning	2%	10% *· ᵇᵉ%*	25%	33% *+ ᵇᵛ ᵉ/ᶜ*
Afternoon	10%	18% *· ᵞᶜ*	15%	19% *+ ᵢᵢ*
Evening	71%	73%	7%	13%·
All Day	72%	76%	34%	42%
All People - Aᵥᵉʳᵃᵍᵉ				
Morning	1%	5%	10%	15%
Afternoon	7%	15%	6%	9%
Evening	68%	70%	3% *5,3*	8% *7.9*
All Day	70%	72%	16%	22%

Figure 10 Videotown 1952: 'TV Viewing and Radio Listening', Cunningham & Walsh.

The study noted that 'Although TV viewing increased over the last year in both morning and afternoon, it is still only about 1/3 as high as radio listening during the morning.'[55] It found that morning time radio could be listened to alongside the undertaking of housework and that 'One of the problems facing TV is that of making daytime programs as interesting to hear as they are to see.'[56] This echoed the wider industry concerns that radio was more accommodating to housewives engaged in domestic work than television could be. If radio remained the best way to address the housewife, then advertisers were less likely to expand into daytime television advertising. As was the case in previous years, the report found that wives and 'older adult females' watched television in higher numbers and for longer periods than men.[57] Despite this, the study emphasized the increases in women's morning and afternoon television viewing which happened in far lower numbers than in the evening. However, the report gave the impression of increasing daytime viewing figures because the viewing figures were slightly higher than the previous year.[58] In other words, the study was pointing to opportunities available to advertisers and broadcasters in the daytime schedule and worked to convince the study's readers that women were a potential audience for daytime television. The daytime female audience, therefore, did not exist (at least in terms of significant numbers); it needed to be created.

The emphasis on the daytime female audience continued in the 1953 and 1954 reports which detailed the slight year-on-year increases in daytime viewing by women. The 1953 study claimed that 'Mother' watched more than 'Dad' and that young women watched more than young men.[59] Mother – who was 'home oftener' – also watched more hours in the weekday evenings (at fourteen hours) than Father (at thirteen).[60] The study acknowledged that evening viewing remained more popular than daytime viewing but noted that the latter was increasing.[61] Both the 1953 and 1954 studies foregrounded the marginal increases in wives' daytime viewing much more so than, for example, children or men's viewing. The focus on the incremental increases in wives' morning and afternoon viewing, which remained miniscule as compared to their evening viewing, gave the impression of a market ready to expand and ready to be tapped. Housewives, the 1954 study found, showed 'a substantial increase in viewing, both in the morning (22 per cent in 1954 and 12 per cent in 1953) and in the afternoon (25 per cent in 1954 and 19 per cent in 1953)'.[62] However, this was not dramatically higher than the afternoon viewing figures for the total audience. By 1955, viewing figures had, in fact, dropped with women watching a lot less daytime television than in previous years.[63]

However, this was downplayed in the report which, instead, maintained the perception of women as avid daytime television viewers. Reports continued to measure the daytime audience through the categories of the 'housewife' and 'all viewers' suggesting an investment in her as the primary audience. Although the report attributed this reduction to the previous year's 'intense interest in the … Army-McCarthy hearings', it nonetheless represented a significant drop in daytime viewing among the audience.[64] Again, as in previous years, what could ultimately be ascertained was that women were avid evening viewers and infrequent daytime viewers. For daytime

HOURS OF TV VIEWING AND RADIO LISTENING IN TV HOMES

	TV Viewing					
	Hours Per Day When Watching			Hours Per Week* All People		
Wives	'53	'54	'55	'53	'54	'55
Morning	2.19	2.16	1.23	1.28	2.35	1.05
Afternoon	2.04	2.18	1.74	1.94	2.70	2.35
Evening	3.56	3.51	3.45	13.90	13.80	15.40
All Day	3.84	4.11	3.89	17.12	18.85	18.80
All People - Average						
Morning	1.98	2.16	1.70	.79	1.50	.95
Afternoon	1.62	2.08	1.45	1.13	1.65	1.40
Evening	3.29	3.16	3.11	12.00	11.70	13.20
All Day	3.52	3.58	3.39	13.92	14.85	15.55

Figure 11 Videotown 1955: 'Hours of TV Viewing and Radio Listening in TV Homes', Cunningham & Walsh, p. 18.

viewing, housewives watched in slightly lesser numbers than in previous years and in slightly higher numbers than the total audience, but not by a huge margin (27 per cent housewives as compared to 20 per cent of the total audience for afternoon programmes, for example). More significantly, the hours of television watched by housewives in the morning was lower than that of the total audience: for mornings, 1.23 hours compared with 1.7 respectively (see Figure 11). This increased throughout the day and, by evening time, housewives viewed for longer than everyone else in the household.[65]

The 1956 report showed that 'for the first time in history, television viewing has reached an apparent plateau – with a decline in both the number of people watching and the hours of individual attention to the set'.[66] The 1957 report showed a similar pattern. In particular, in 1956 the percentages of all viewers (including men, children and other female adults) and housewives watching daytime television edged closer together such that there were only small differences in the numbers viewing during the day. However, the report continued to compare the viewing of the housewife and the total audience. While there was a slight increase in the overall hours watched by women throughout the day, there was a decline in the percentage of women watching such that the women's viewing was almost equal to that of the total audience for the morning and afternoon. The 1956 report also indicated that women were disinclined to watch television while doing household chores. Only 8 per cent did chores while watching, in comparison to 26 per cent of women listening to radio while engaged in household work.[67] Despite the best efforts of broadcasters, advertisers and audience researchers, housewives' daytime television viewing was more the exception than the rule. Nonetheless, audience researchers who carried out these studies continued to concentrate on the housewife when studying the daytime audience. While the housewife did sometimes watch in slightly higher numbers than others at certain times of the day, this was no different a pattern than for evening viewing, where she watched more than others. However, much less emphasis was placed on the housewife's evening viewing than on their daytime viewing.

By 1958 – the last year of the Videotown study – wives' viewing throughout the daytime did not significantly differ from the viewing of all people, even though the studies continued to emphasize her viewing practices more than anyone else in the family.[68] For example, although wives watched somewhat more daytime television between 1954 and 1955, by 1958 the percentage of wives' viewing compared to the total audience for morning, afternoon and evening was as follows: 11 per cent compared to 12 per cent, 25 per cent compared to 23 per cent, and 87 per cent compared to 79 per cent. In other words, the percentage of wives viewing in the morning and afternoon was not so different than all other viewers, despite the impression that her viewing patterns were somehow unique. From 1956 to 1958, wives' daytime viewing was similar to that of the general audience. If wives showed any difference to the general audience, it was that they were much more inclined towards evening viewing. This was also reflected in the hours spent watching television. Although women watched for less hours than the total audience in the morning, they watched for marginally more hours in the afternoon and quite a lot longer in the evening.[69] The Videotown study ultimately demonstrated, therefore, that the introduction of gendered data was less the result of clear patterns of gendered viewing and more an institutional imperative based on 'common sense' assumptions of how people (specifically, the housewife) might watch television. Despite demonstrating that women watched evening television in higher numbers and for longer hours than the total audience, the studies concentrated more on her daytime viewing since this was what broadcasters and advertisers were increasingly interested in. And despite the fact the women watched in far less numbers and for far fewer hours in the daytime than in the evening, the studies insisted on continuing to focus on her viewing more so than other audiences during the daytime hours. In its persistent pursuit of the housewife through audience research, the television industry effectively cultivated rather than discovered a daytime audience of housewives.

Along with the launch of the Videotown studies, further audience research studies sought to define and make sense of the television public. While the Videotown studies took a number of years to turn the spotlight on both the daytime audience and housewives, other audience research companies were quick to pay close attention to the emergence of daytime television and to measure and study its audience. Advertest was one such company. Specializing in market research on radio, television and film, Advertest produced research on set ownership, advertising effectiveness and programme preferences. During the late 1940s and throughout the 1950s, Advertest conducted a number of studies that focused exclusively on daytime television viewing and the figure of the housewife viewer. From as early as 1949, Advertest published regular monthly reports on trends in television ownership and television viewing including daytime television viewing. Advertest was often commissioned by broadcasters such as NBC and advertising publications such as *Sponsor* to produce research on the television audience. Its studies ranged in subject from broad surveys of evening television viewing to studies of television genre viewing and the effectiveness of advertising. The results of Advertest's reports were summarized and published in magazines and periodicals such as *Broadcasting* and *Sponsor* and were used to track changing habits among television audiences. Beginning in March 1949, and under the series title 'The Television Audience of Today', Advertest published its first report on daytime television, the 'Study of Daytime Television'.[70] A follow-up study of daytime television audiences was published in 1950.[71] The two reports, discussed here, established the housewife as the main viewer of daytime television and concentrated mainly on her viewing practices. Not only did the studies use sample groups primarily composed of housewives, but they gave the impression that increased viewing of television across the years 1940 to 1950 was specifically due to increases in housewives' viewing when, in fact, it was in many ways a result of more viewing by other members of the family. The close attention paid to the housewife – identified as the main viewer because it was assumed that she was home – meant that the studies supported the 'common sense' claim that she was the inevitable audience for daytime television.

The 1949 study, for example, did not make immediately clear that it had surveyed women only. This only became apparent in the lack of data on viewers other than children or wives during the morning and early afternoon periods. This gave the impression that only wives and children watched during the day and that no men watched in the early morning and up until 3.00 pm. In addition, many of the questions were directed specifically towards the housewife. For example, housewives were asked what time of day it was convenient for them to watch television.[72] They were asked what programmes they preferred[73] and what prompted them to watch at certain time.[74] The study asked when the television set was first turned on in the day as well as who suggested it be turned on. Overall, the results were that children were most likely to suggest turning on the set during the day (14.9 per cent). Wives were second at 12.4 per cent and husbands were least likely at 2.2 per cent (see Figure 12). However, it also found that 'in 19% of the homes using sets before 5.00 pm more than one person usually watches daytime television'.[75] Therefore, although housewives were considered the primary audience

```
                    TOTALS

                DAYTIME VIEWING


The total percentage of sets that are usually used
before 5 pm (total of four previously mentioned
periods) is

                    29.5%

Out of this number the totals for each group respons-
ible for first suggesting that the set be turned on
are as follows.


        Children            14.9%

        Wife                12.4%

        Husband             2.2%
```

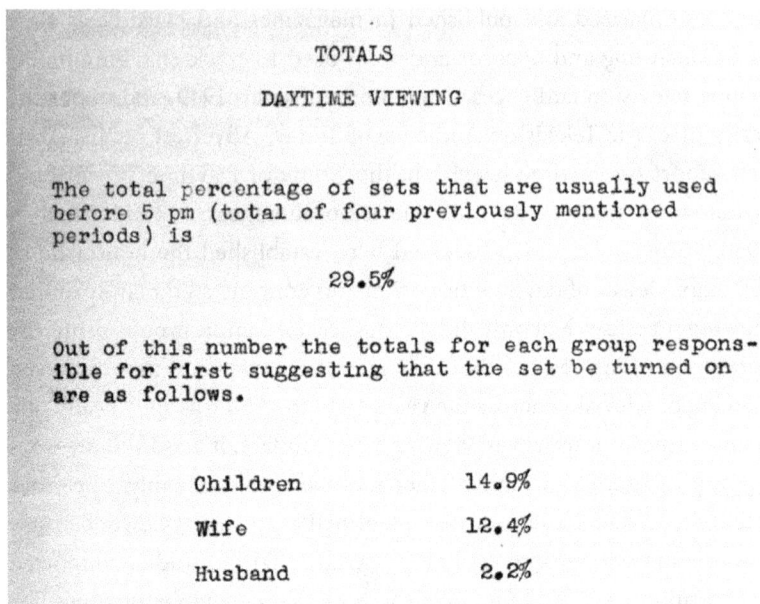

Figure 12　Advertest 'Study of Daytime Television', 1949, p. 7.

of daytime television, it was instead mainly children who initiated television viewing, with multiple family members viewing when the set was on.

The study measured when viewers began watching throughout the day and noted that viewing was concentrated in the evening time and was over double the viewership than at any point throughout the day. Figures for daytime viewing were low. Of the 8.5 per cent of respondents who watched before 10.00 am, children were most likely to request that the television was turned on (5.9 per cent) with only 2.6 per cent of wives doing so. However, the study noted that wives would watch along with children.[76] It suggested that few were viewing in the hours between 10.00 am and 12.00 pm and attributed this to 'a number of factors including (1) the housewife preparing lunch and (2) the absence of any highly rated programs'.[77] This would seem to suggest that housewives had neither time for, nor the interest in, watching television during these hours. Advertest found that viewing figures increased between 12.00 pm and 3.00 pm and that 'it is in this period that the wife first takes over as the user of the set. This can be explained by the heavy concentration of women's programs at this time'.[78] However, a high number of other viewers were also found to be watching during this time. It was simply the case that the husband was not included in the data on morning viewing. His viewing was measured from 3.00 pm onwards only. The study found that more viewers turned on the set between 3.00 pm and 5.00 pm than at any other time during the day. It attributed this to '(1) the return of many children from school, (2) the arrival home of the husband and (3) a free time period for the wife'.[79] During this time period, a total of 9.5 per cent of people first turned on their television sets: 4.1 per cent of wives, 3.2 per cent of children and

```
THOSE WHO TURN ON THEIR SETS BETWEEN 3 AND 5 PM


        Total in this group      9.5%


During this period we find that the greatest number
of sets are turned on for the first time (of any day-
time period.) Factors responsible for this include (1)
the return of many children from school, (2) the
arrival home of the husband and (3) a free time period
for the wife. We also find (as in the previous period)
an abundance of popular programs being presented at
this time.


"AT WHOSE SUGGESTION (FOR WHOSE USE) IS THE SET
    FIRST TURNED ON AT THIS TIME?"


        Children              3.2%

        Wife                  4.1%

        Husband               2.2%


For the first time we note that the husband enters
the picture as a user of the set.
```

Figure 13 Advertest 'Study of Daytime Television', 1949, p. 8.

2.2 per cent of husbands (see Figure 13). Advertest found that this time period had the highest number of multiple viewers.

Ultimately, the data indicated that daytime television viewing was much less popular than evening television and that television viewing increased when there were more family members in the household. Children were the most likely to turn on television, particularly before 12.00 pm. Wives were most likely to turn on the set between 12.00 pm and 5.00 pm. However, even in these time periods, the numbers turning on their sets for the first time never passed 5 per cent of the total audience. While this did not reflect the total number of people watching (since people may have turned on the television set earlier in the day and continued watching), it nonetheless suggested that daytime television had quite a small audience even among housewives. However, the framing of the figures on daytime viewing masked this fact.

Advertest also discovered that housewives would not watch television simply because it was available. Nor was the promise of women's programmes enough to attract housewives to television. When non-viewing housewives were interviewed, 43.9 per cent of them cited having 'no time' as their main reason.[80] And 14.3 per cent said they were simply not interested and 6.6 per cent of them said they were not usually at home. In addition, when respondents were asked what times were most suitable for viewing, it seemed that the daytime was not very convenient at all. While 10.4 per cent of housewives suggested that the period 3.00 pm to 5.00 pm was most convenient, figures for the remaining hours of the day were low, with less than 5 per cent stating a preference for most other time slots throughout the daytime.[81] The report also noted that housewives were selective about what programmes they watched. The housewife would make time to see programmes she liked, even if the time was inconvenient. On the other hand, Advertest found that housewives would not watch any television programme simply because it was convenient to do so at a certain time.[82] Among the more popular programmes among those surveyed were variety and feature films, with over 5 per cent of respondents reporting regular viewing. Women's programmes were also popular, with *Okay, Mother*,[83] *Needle Shop*[84] and *Television Shopper*[85] cited in the top ten. When asked what other programmes they would like to see on daytime television, respondents showed strong preferences for 'more and better films', variety and educational programmes.[86] Housewives, it seemed, sought a much wider range of programmes and genres than was available.

A further study published in June 1950 was largely concerned with the numbers of viewers for particular time periods throughout the day. However, in somewhat of a contradiction, the study indicated that 'whenever possible, the housewife was questioned personally since it was felt that she would be the major user of the television set before the hour of 5.00 pm'.[87] Although the study aimed to offer 'a picture of those who view Daytime Television', it nonetheless opted to restrict this picture to those who were predetermined to be the primary audience (despite the fact that the previous year's study indicated that children were the primary daytime audience). The report suggested that, by 1950, daytime audiences had increased to 44 per cent, up from 29.5 per cent the previous year.[88] Advertest's data on 'the time of day the television set is first turned on' by regular viewers showed that 'female adults' were by far most likely to turn on the set before 12.00 pm, with no 'adult males' and some children doing so. By midday, however, children were most likely to turn on the television set, followed by female adults and then adult males. While more women reported being the first to turn on the television set overall, there were periods where children did so in much higher numbers and there was significant representation by adult males, particularly in the late afternoon (see Figure 14).

This suggested a much more diverse daytime audience which contradicted the study's claim that housewives were the major users of television. When respondents were asked whether they rearranged their daily household chores to watch television, 87.2 per cent said that they had not rearranged it in any way.[89] Like the previous year's report, then, housewives would only watch television when it suited them to do so. In terms of programme preferences, the favourite cited programme for regular and occasional viewers was the Western film

SET FIRST TURNED ON	PERCENTAGE	RESPONSIBLE FOR FIRST TURNING SET ON Male Adult	Female Adult	Children
Before 9:59am	4.3%	-	91.7%	8.3%
10:00-11:59am	14.9%	-	73.8%	40.5%*
12:00-2:59pm	31.9%	32.2%	45.6%	60.0%*
3:00-4:59pm	48.9%	9.4%	26.8%	74.6%*

Figure 14 Advertest 'Study of Daytime Television Number 2', 1950, p. 4.

aired daily at 4.00 pm.[90] Feature films were popular among respondents, as were programmes aimed at women and housewives as well as sports programmes.[91] When asked if they turned on television to watch anything at all or to watch something specific, 85.1 per cent of respondents said they turned on to view a particular programme. And 22.7 per cent[92] reported this as a Western film, 6.4 per cent as variety programme *Okay, Mother* and 6 per cent as variety programme *Rumpus Room*.[93] In other words, Westerns were the most popular and the most likely programme to encourage appointment viewing. Given the tendency to assume that women were the primary viewers of daytime television and that women tuned in for women's programmes, this data seemed at odds with the common sense assumptions about the female audience, who were thought to make time for women's programme.

The Advertest studies, with their emphasis on the housewife, contributed towards the discourse of the housewife as the key audience for daytime television. However, the Advertest data revealed that television viewing was a family affair, with many family members partaking throughout the day. Although women tended to watch at various times throughout the day, the Advertest studies indicated that children and men also watched in significant numbers, especially in the afternoon. Yet this was not necessarily how Advertest interpreted and represented its findings. Nor was it how others understood and represented the reports. Those advertising and broadcasting publications that featured the Advertest daytime studies continued to correlate daytime television with the housewife audience, such that further features on the daytime audience and summaries of Advertest reports referred almost exclusively to the housewife when discussing the daytime audience. A 1952 Advertest survey carried out for *Sponsor*, for example, and like previous Advertest studies, sampled only women about their 'televiewing before 12 noon'.[94] A 1 April 1952 issue of *Sponsor* published the findings of an Advertest report that once again asked 'female adults in TV homes' about their preferred weekday for daytime viewing.[95] Yet another Advertest survey, once more carried out for *Sponsor*, asked only women if they listened to or watched *Arthur Godfrey and His Friends*[96] in the mornings.[97] This effectively rendered female viewers important only in relation to their daytime television consumption, with the result that the television schedule was organized around the gender of the viewer.

Daytime television was the terrain of women, regardless of the fact that they viewed the most in the evening. Evening, prime-time television became, as Meehan argues, the default terrain of '*the* audience', which Meehan suggests was gendered as male.[98] The daytime audience, thus, became synonymous with the housewife in part because of the methods deployed by audience research organizations as well as the preconceptions that audience researchers brought to their assessment of the daytime audience. The studies and reports were used to determine that there was a valuable audience for daytime television, even if not nearly as valuable as the evening audience, and to draw advertising spending for daytime programmes. By identifying a clear and distinct audience in housewives, broadcasters and audience researchers could produce a coherent message about daytime television and its audience. In the context of daytime television, the audience that held value was the housewife, even though she formed part of a larger viewing public of children and men. Throughout the 1950s, broadcasters and audience researchers continued to seek out the 'lady of the house' when collecting data on daytime viewing. For the most part, this was sufficient since the project was less to produce a comprehensive analysis of daytime viewing than to provide advertisers and programme-makers with data that gave them confidence in the networks' capacity to attract audiences. This 'commodity audience' of housewives was produced through in-depth studies of the housewife's socio-economic status, lifestyle, purchases and habits. However, the emphasis on the housewife in audience research studies and reports, at times, meant that little attention was paid to other viewers who didn't fit this profile. And when there was interest from brands or advertising agencies in the wider population of daytime viewers, the data simply wasn't there.

This was evident, for example, in the NBC Research and Planning Department's study *Television's Daytime Profile: Buying Habits and Characteristics of the Audience*, which was carried out by W.R. Simmons & Associates in January 1954.[99] The results of the study were then issued in reports and bulletins personalized for advertisers and brands who requested it. Although the title suggested that the study was based on the total audience (e.g. a sample of all viewers who may be watching), it was, like the Videotown and Advertest studies of daytime audiences, inclusive of women only. This focus on the housewife occasionally worked against NBC when clients and advertisers requested data on other members of the household. The study was based on 'personal interviews with 3,243 women living in 2,871 households'.[100] That the survey only included data on women reveals the extent to which women were perceived – within NBC as elsewhere – as the main audience of daytime television. The survey also pointed to the NBC Research and Planning Department's concern with the consumption patterns and status of its female audience. Not only did it provide data on the age, wealth, number of children and educational level of the household, it also provided rich data on the purchases in these households that viewed programmes such as *Today*,[101] *Home*,[102] *Ding Dong School*,[103] *The Pinky Lee Show*[104] as well as many other daytime programmes. This was used to communicate to clients the success of the programmes and the advertising effectiveness of television. Indeed, as Cassidy, Spigel and Stole note, a number of NBC's daytime programmes had a distinctly consumption-centred agenda and structure.[105] Magazine style programme *Home*, for example, was primarily designed to sell select products to the audience and although it had a number of other segments related to news, culture and politics, it worked to integrate a consumerist sensibility in its mode of address, one aimed squarely at the housewife.[106]

The importance of this mode of address and the housewife's reception of it was clear from the audience studies carried out by the NBC Research and Planning Department, many of which offered elaborate detail on the spending patterns and power of its female daytime viewers. This often included data on the annual family income in the housewives' household, the occupation of the head of the household, the figures for ownership of durables, entertainment devices, home ownership and car ownership. It also included data on the age, educational level and marital status of the 'lady of the house' along with her employment status, data on topics of interest to her, whether she smoked, if she had a vacation, her shopping habits including items bought as well as when she shopped. NBC presented this data to clients in numerous reports and letters that aimed to demonstrate the high-quality audience of medium- or high-income housewives it had for its programmes. For example, a report on *Today's* daytime audience included data on weekly expenditure for groceries and indicated that those who spent over 20 dollars a week were somewhat more likely to be *Today* viewers in comparison to non-viewers.[107] The report also indicated that the educational level of *Today's* female audience was generally higher than that of non-viewers, with 51.3 per cent being education to high school level or above as compared with 45.5 per cent of non-viewers.[108] All of this worked to suggest that the *Today* viewers represented a high-quality market with spending power and status.

Yet, while NBC had indeed produced extensive studies on its daytime audience of housewives, the absence of other audience segments did not go unnoticed and, at times, the NBC Research and Planning Department had to explain the lack of wider data on the daytime audience of, for example, children and men. In a 24 September 1954 letter from NBC Research and Planning Department to its client Royal

Typewriter regarding the *Today* audience data, NBC acknowledged that it only had data for women and not the total audience.[109] This suggests that although NBC had prioritized data on women only, clients were seeking information on all daytime viewers. NBC tried to work around this gap by explaining that:

> Our study was conducted with women only, and the viewing data which we obtained was for these women – men and children were not questioned on any aspect of the study. We did obtain, however, the occupation of the man of the house from the women … Looking at the last two columns [provided in the report], you can see that TODAY homes have an edge in the Professional category (23.6 % vs. 18.6%) and in the Craftsmen category (44.6% vs. 34.5%).[110]

NBC went so far to suggest that it had also checked with Advertest for other data on other viewers but that Advertest, too, did not have any such data.

This issue was also apparent in a 21st December 1955 letter from the NBC Research and Planning Department to an NBC sales executive seeking information on the occupations of the audience for a manufacturing client.[111] NBC again attempted to extrapolate data on the occupation of the male head of household through the data it had from female respondents. It was apologetic for not having direct data on the occupation of *Today* viewers since it only had data on the female householders and their responses to questions about the occupation of the head of household. NBC stated that:

> This is the only information on this subject which we have at the present time. Perhaps your manufacturer will be satisfied with this substitute; I believe it still gets across the fact that an 'executive-type' audience probably watched the program.[112]

In other words, where NBC had assumed that the only data worth noting was that of female daytime viewers, it found that clients wanted data on all viewers – data that NBC did not have. As it noted in a 25 January 1956 letter regarding the programme *Meet the Press*[113] 'unfortunately the study [Television's Daytime Profile] did not include similar data on male viewers'.[114]

That audience researchers spent little time accounting for the male audience is telling. In early commercial television audience research of the 1940s and 1950s, the gendering of the audience occurred primarily through the figure of the housewife and in relation to daytime television. Thus, women became the bearers of gender in a way that men did not. The female audience, as a subcategory of the audience, was further down the hierarchy of taste and status. This had implications for the way that women were addressed by television. Michele Hilmes suggests that the early US broadcasting industry developed a programming strategy that created a

> differentiation between daytime and nighttime programming, by which daytime became the venue for a debased kind of commercialized, feminized mass culture- heavily dominated by advertising agencies – in contrast to the more sophisticated, respectable, and masculine-characterized arena of prime time.[115]

US television audience research worked to embed this gendered logic through its studies and reports which cast the housewife audience as a daytime-viewing, consumerist and middle-class one that appealed to advertisers.

By the mid-1950s, the daytime audience became synonymous with the housewife. Programme planning and advertising strategies followed this logic. This relationship between the broadcasting industry, commercial audience research companies and the advertising industry facilitated the development of the concept of the female audience, one that was defined not only by gender but also by race, class and social status. Commercial audience research produced this female audience, broadcasters produced programmes for her and the advertising industry produced advertising and sponsorship that encouraged her to consume more products. Daytime television proved a suitable vehicle for this since, as audience research seemed to suggest, the housewife was ready and waiting to be addressed during these hours. Following on from this, and as Kristen Hatch explains, 'by defining daytime programs as appealing to housewives, sponsors … could fine-tune their appeals to women without alienating a "general" (implicitly male) audience that might balk at being identified with this disparaged category of viewers'.[116] Thus, US audience research, and the broadcasting industry more generally, worked to reproduce and perpetuate the division of gender that promoted a feminine ideal that was white, middle-class and domestic. Actual women were, therefore, only relevant insofar as they could identify with this ideal and, as such, form part of the female audience. Nor was the United States exclusive in such a definition of the female audience. In Britain, the development of BBC television audience research resulted in a similar framing of the female audience. Unlike the US broadcasting industry, British television broadcasting was, in the first half of the twentieth century, a public service and, therefore, had no immediate requirement to appeal to advertisers. Yet the profile of the female audience that emerged was equally as problematic. In the final chapter, I discuss the audience research undertaken at the BBC between the 1930s and the 1950s, during which time the female audience came to be understood as a domestic and daytime one.

The British female television audience

Audience research in Britain took longer to institutionalize than in the United States, in part, because British broadcasting did not have the same commercial pressures and did not have to rely on revenue from advertising or sponsorship. While the US radio broadcasting industry very quickly saw audience research as crucial to its economic security, the BBC was initially quite resistant to what many in the organization saw as pandering to the masses. As described by BBC Head of Audience Research during these years, Robert Silvey, there were those 'who simply refused to believe that any systemic study of the public was ballyhoo; its growing use revealed the cynicism – or gullibility – of business men'.[1] He noted that the issue was divisive, however, with others arguing that the BBC could hardly be said to be a public service if it didn't engage with the public. Up until the 1930s, the BBC chose to provide what it thought was in the public interest not what the public said it wanted. However, this position was increasingly challenged. According to Sean Street, print media had begun to complain about the BBC's radio broadcasts and encouraged the public to do the same.[2] More importantly, the BBC faced competition from pirate radio which targeted listeners with more entertaining programmes.[3] Thus, the BBC came to accept audience research much later than US radio broadcasters, and more reluctantly. The Listener Research Department was only established in 1936 but was not fully embraced within the organization until after the war and when television was emerging as a rival to radio. The audience research undertaken was intended to be quite different to that of the United States and to emphasize the sociological above the statistical.[4] Focus would be given to the engagement of the audience rather than simply audience figures.[5] As Shaun Moores argues, one of the main features of BBC audience research was its 'investigations into the temporal rhythms of the day-to-day' as interpreted in relation to categories such as gender, age and class.[6] As I argue in this chapter, the audience research that emerged produced an audience segregated by gender and time of engagement, with the result that women were – as in US audience research – understood almost exclusively as daytime listeners and viewers. I suggest that through its audience research, the BBC ultimately gave its female viewers not what they wanted but what the BBC thought its female audience wanted. These were not necessarily the same thing.

This points to the vast difference between the actual audience, composed of complex individuals, and the institutional audience manufactured by audience researchers. This difference has been noted by Ien Ang who claims that:

> Quite obviously, before there was television, there was no such thing as a television audience. The television audience then was not an ontological given, but a socially-constituted and institutionally-produced category.[7]

It is possible to see how the female audience became such an institutionally produced category at the BBC through analysis of its audience research carried out from 1936 onwards. The BBC first began collecting and studying data on its audiences with the establishment of the Listener Research Department (later, in 1950, Audience Research Department). The department deployed methods and techniques of measurement and classification that claimed truths and produced facts about the female viewing public. The tools and techniques used to assess the viewing public – shaping them into the 'audience' – were regarded as objective and scientific.[8] Thus, the audience produced by such research was rendered as a quantifiable 'fact'.

Nonetheless, this supposed scientific objectivity at the BBC relied on social discourses of identity, as well as on behavioural norms. When collecting information on the audience, the BBC interpreted it according to selected demographic classifications such as gender, class and age, and so the public became pre-defined according to certain criteria as opposed to others (i.e. gender rather than height). In addition, early BBC television audience measurement drew its interpretations about the audience from samples that were made up of those in possession of a television set, in other words, those who could afford this expensive new device. Thus, the knowledge that the BBC (initially at least) produced about its female audience was based on quite a narrow segment of the female population. Those early methods of collating data, in turn, informed how the BBC undertook television audience research in the post-war years, particularly in regards to its emphasis on the daytime female audience. According to the BBC, the female audience was imagined as a daytime one. However, it was exclusive of those working or otherwise occupied women who formed a significant part of the potential audience. The female audience, then, was less a fact and more a product of gendered institutional practices within audience research.

The BBC's production of the female audience, as well as its use of such a category in programming and scheduling practices, points to some of the issues that have been raised about the concept of the audience more generally and the means by which it is put to work. Lisa Gitelman, for example, suggests that data is never 'raw' or neutral; rather it is 'cooked'.[9] The process of segmenting the public according to gender already imposes a certain interpretive logic on the data. In addition, the more large-scale and quantitative the measurement used, the more removed from social reality this institutional 'knowledge' becomes. In other words, when knowledge is formed through the use of the categories of 'audience' as well as such classifications, as for example, 'male' or 'housewife', what is lost are the complexities, nuances and personhood of individuals. Philip M. Napoli confirms, 'efforts to enhance knowledge, predictability, and control in relation to the audience have ... been accompanied by the kinds of analytical simplification that have historically been associated with the process of rationalization'.[10] For example, when audience research and measurement begin with the collection of data on women, it already assumes shared characteristics in this group of the population resulting in preconceived correlations between their viewing habits. It may seem inevitable that the viewing public might be understood and acted upon in these terms. Since gender is one of the key markers of identity in society, the prevalence of these categories has materialized through decades of refinement of audience research methods.

We can note from the early, experimental years of television – between 1936 and 1939 when the war resulted in the cessation of broadcasts – that there was initially at least a distinct separation of audience research and programme planning and production. In other words, very early audience research did not have a direct impact on what was produced and scheduled for the 'audience' as constructed by the audience research department. Between 1936 and 1950 – at which point audience researchers began producing audience research reports on individual episodes of programmes – there was a more informal relationship between programme production departments and the audience research department. Indeed, Head of Listener Research at the BBC, Robert Silvey, had maintained that such a distinction was necessary:

> From the very beginning of my time with the BBC, I constantly stressed that audience research's function was limited to providing the decision-makers with information upon which they could act – or not act – as seemed to them right. Map-making and navigating were quite different functions. Ours was map-making.[11]

By the 1950s, there was more communication and interaction between audience research departments and programme production departments as evident in the production of viewer reports that were issued to programme-makers from the 1950s onwards. And, later still, the relationship between audience measurement, programme planning and production became closer but was still not entirely integrated into an overall strategy for acting upon the findings of audience research reports, surveys or viewing reports. For example, if viewership of programme X fell below figure Y, then the programme would be cancelled or rescheduled. The case of BBC television audience research enables us to understand how the classification of the viewing public took root and was increasingly embedded in the institutional logic and practices of the organization, firstly focusing on radio and later television.

Looking back at archive materials from the 1930s through to the 1950s, it is possible to identify how and where gender became integrated into the knowledge production mechanisms of television audience research. Over time, as this knowledge was disseminated to various other planning and production departments, the 'female audience', 'the female viewer', 'the housewife' and 'the mother' come to represent what was imagined about, and inferred from, the viewing public. Early television audience research did not make sense of the audience in gendered terms, even though radio research was doing so. Instead, the measurement of gender was later introduced to television and this knowledge was put to use across the organization, for example, in the development of women's programmes, the scheduling of these programmes and the partitioning of programmes and audiences along gendered lines. The analysis of the audience research carried out at the BBC reveals that audience research did not 'discover' that viewing habits, tastes and experiences were determined by gender. Rather the classification of gender entered into the lexicon of audience research reports and, as a consequence, extended into the institution more widely.

The development of audience research at the BBC from 1936 was, as Stefan Schwarzkopf notes, undertaken in response to press criticisms that the organization was not representing the public interest and thus was not providing a valued service to its viewers.[12] During the 1930s and early 1940s, the infant television industry was more inward-looking, working as it was to develop a form of broadcast that might suit the new medium. Any sporadic engagements with the audience were more for the purpose of using viewers as early testers of the medium. Feedback, often in the form of letters, enabled programme-makers to adapt or improve their broadcasts and acted as a barometer of tastes and interests.[13] As such, early attempts at audience research were less concerned with the profile of viewers and more with the reception of individual programmes and programme formats, as well as the general enthusiasm (or lack thereof) for television. The BBC, however, was keen to encourage viewer feedback. Through its publication, *Radio Times*, the BBC engaged with its audience by responding to individual reactions from the public and acknowledging the receipt of letters and phone calls, as noted in an issue of the magazine from 15 April 1938. An interest in the composition of the audience would only materialize once television was viable as a mass medium and a business in the post-war years. In the pre-war years of the television service, any audience research undertaken, whether formally or informally placed emphasis less on who the audience was and more on how the television service was received. Consequently, there was little reference to women as a category, market or social class of viewer. If daytime talks were popular in the afternoon, this was not immediately correlated with the female viewer, unlike in the post-war years where the daytime schedule was determined as the primary terrain for 'women's programmes'.

When BBC television broadcasters began to formalize audience measurement in the post-war years, women emerged as a distinct group among the general audience. Subsequently, the female audience became isolated from the general audience. In other words, where programmes were viewed equally among men and women, they were understood in terms of genre. Where programmes were preferred by more women than men, they became 'women's programmes'. Equally, when men and women watched at the same time, schedules contained programmes that were of interest to the general audience. However, when women watched at certain times more than men, this schedule came to be dominated by women's programmes (instead of general interest programmes that women might have also enjoyed) and regardless of whether women enjoyed such programmes. While these might seem logical from the broadcaster's point of view, it had the effect of partitioning women's programmes from the broader schedule. It also produced a narrow understanding of the female audience, one which was interpreted as domestic and maternal, despite the fact that many single or married working women did not fit this category since they were employed outside the home. In 1947, for example, 6 million women were in the workforce and in 1951 women made up 30 per cent of the full-time labour market.[14] Ultimately, the female viewer represented an institutionalized form of knowledge about how the viewing public was composed. This was based less on the real conditions of women's domestic and non-domestic lives but on the social conventions and norms that held women as domestic and middle-class. Such ideas about women formed the lens through which female viewers and data collected on them were interpreted. This had implications for what television would become, including how it was scheduled, what types of programme and commercial content was broadcast and how the general public was addressed. Tracing the practices in television and radio audience research from its beginnings in 1936 to its formalization in 1950 shows the extent to which the idea of the female audience was developed through a series of problematic surveying and somewhat inaccurate representation of the female viewing public.

During the experimental years of television any interest in, and reference to, the television viewing public was largely related to viewer responses to the quality of production and broadcast as well as image and sound. At this early stage of the television service, so little was known about the viewer that there was no attempt at segmenting an already small audience into categories such as male and female. During this time, the BBC concerned itself more with the improvement of the technical operations of television broadcast and the development of television programmes. As the service developed, there was some concern about how the audience would respond to television, and television was imagined as at the mercy of the public's tastes. It must be remembered that during the early experimental years only approximately one hundred households owned television sets, only increasing in fairly conservative numbers between 1936 and 1939.[15] Private set ownership remained extremely low until the early 1950s.[16] The BBC was still concerned with the audience as it was eager to encourage more television sales and to develop the service. A Postmaster General's report from January 1935, for example, referred to television being 'put to the acid test of public opinion'.[17] In an August 1936 report of the television service, the public's reaction to television was of utmost importance, and the report suggested that the public would ultimately shape what television became. In a section on the reaction of the viewing public, it noted that:

> This is going to be extremely difficult to determine at first because we have to separate the interest due to novelty from that arising from genuine entertainment. It may be found that methods which have been laid down for ordinary broadcasting may be completely unsuitable.[18]

Here, the public was invested with a certain amount of power, and it was implied that their response to television would guide programme production. This comment also suggested a sense of powerlessness on the part of the BBC due to its lack of knowledge about its audience. Such early concerns with the opinion of the viewers led to one of the first efforts to carry out audience research on the responses to television.

In December 1936, the BBC put out a call to its small viewing public to engage them in a survey about its programmes. It received seventy-four completed responses, though the sample was largely composed of those who were already working or otherwise involved in broadcasting.[19] Although carried out within the newly formed Listener Research Department, the Viewers and the Television Service study – published in 1937 – was more a survey of television in the public than who the television public was. Nonetheless, the report provides some insight into how the BBC made sense of its audience. The report was based on a sample of respondents who were affluent enough to own a television set since the cost at this time was approximately 6,120 guineas.[20] This left it out of reach for those earning an annual wage of approximately 200 pounds.[21] The resulting survey findings, then, were based on a narrow representation of the general public. The objectives of the survey included the following:

1. To find out how many private viewing sets were in the hands of the public;

2. To find out under what conditions the television programmes were being received;

3. To find out viewers' opinions on the television programmes;

4. To find out the number of places where the television sets were installed for the purposes of demonstrating the service to the general public.[22]

The report did not ask specific questions about individual programmes but it did allow space for comments, which was exploited by many of the respondents. Because the comments were quite broad-ranging, it was difficult for those interpreting them to produce any meaningful conclusions. In fact, their report noted that this had 'its disadvantages as a method of obtaining clear guidance about the views of the television programmes'.[23] Later research efforts would turn towards more quantitative methods to better manage the data. In the 1937 report, programmes – rather than viewers – started to be categorized. In this report, responses to programmes were quantified in terms of genre and enjoyment. This was accompanied by summaries of general views about the programmes. Of particular note was the viewer feedback on the BBC's instructional and educational programmes often centred on domestic chores, named in the report as Studio Demonstrations and Talks. Such programmes included titles like *Quarter-of-an-Hour-Meals*,[24] *Accidents in the Home*[25] and *Demonstration by the Women's League of Health and Beauty*.[26] Given their subjects and titles, the programmes might be assumed to be – but not yet identified as – 'women's programmes'. The responses to them were largely unfavourable. 'Disapproval concentrated largely upon demonstrations of cooking, washing, ironing, etc., which were condemned as of little interest to those who could afford television sets. It was also pointed out that fashion parades were of little use given the absence of colour'.[27] This gave the impression of an affluent television audience.

Since television ownership at this time was confined to those of sufficient economic means or those with access to the public viewing rooms (where demonstrations of television programmes were provided), it is possible to assume that the demographic of the television audience was more determined by class and social status than by gender. The 1937 survey of seventy-four respondents was made up of both private owners of sets and those set owners who operated viewing rooms. Although the number of private sets was likely to be much higher than the number of survey respondents, it was far from being as ubiquitous as radio. Estimates for British pre-war television set sales were 20,000.[28] These television owners and viewers became understood as the television audience by the BBC. In effect, then, the BBC's first efforts at programmes were influenced by some general stereotypes about female interests including domestic chores, fashion and beauty, while its post-1937 survey impression of the female audience was equally narrow: in this case limited to the affluent middle-classes.

Further television audience research continued in 1939 with a number of ad hoc measures to develop an understanding of the television audience. By that time, the Listener Research Department was already established and, although not a priority among others at the BBC, some further research was carried out on television. In the same year, the BBC held a television conference which invited 150 television viewers to ask questions of the Director of Television, Gerald Cock. The aim of this conference was to offer the public an opportunity to speak with and give feedback to BBC representatives. The two very different forms of interaction with the viewing public offer a good means of representing how and why formal audience research became the dominant means of developing knowledge of the viewing public in the post-war period. The television conference was more an enquiry of television by the public, whereas the Television Enquiry carried out by the Listener Research Department was a study of the viewers of the television service. The television conference was held in June 1939, after the Television Enquiry interim report had been published; however, it represents some of the earlier ways that the viewing public was understood (or not understood, as the case may be).

The conference was intended to provide a forum in which to inform television viewers of how television was developing and to enable them to ask the staff questions about plans for the service. Viewers were also able to make suggestions about what they expected of television. Both men and women raised questions about, and offered feedback on, the variety of programmes, the quality of television and the problems with broadcasts. While the intention of the conference was perhaps to showcase and promote the television service, the transcription of the conference suggests that it was a troublesome affair.[29] Cock was quite defensive and dismissive when asked about the expansion of the service and the possibility of more programmes. In one exchange with a woman, who suggested the production of a 'children's hour', he agreed that this would be a valuable addition to the service but stated that it was not within the television department's means to guarantee that this would happen. In fact, following the conference Cock wrote that future conferences should ban speeches from viewers and lamented that viewers had written to him to complain about his responses.[30] Such an experience perhaps made those involved in the production of television more cautious about directly engaging with audiences and it is, therefore, no surprise that further assessments of audiences took place by those directly engaged in more quantitative audience research.

The Television Enquiry of the same year, on the other hand, shifted focus from individual viewers to the viewing public and it is possible to see the formation of the television audience in the earliest television surveys carried out. Using a sample of nearly 1,200 television set owners, the survey asked respondents a series of questions about what kind of sets they owned, their programme preference, as well as more specific questions about viewing habits. In comparison to later television surveys in 1948 and 1951, this research was again less concerned with developing knowledge of the composition of the audience and more interested in understanding general engagement in the service. While gender remained absent as a category of the viewing public, methods of quantification and segmentation were deployed. The interim report from 4 April 1939 introduced the research method and findings and quantified responses by percentage.[31] In other words, percentages of total responses – as opposed to individual comments – became important ways of understanding audience behaviour. For example, in asking how many people watched television or how many preferred the use of intervals between programmes, the audience was classified in terms of simplified responses (yes/no). Unlike the television conference, there were only limited opportunities for respondents to account for or contextualize their responses. This represented one of the initial efforts to produce the audience using quantitative methods and to turn it into something far more manageable than the unruly viewers that participated in the television conference.

By the time the enquiry was completed and published in June 1939, classifications had entered into the system of measurement.[32] For example, the 865 respondents were classified according to occupation, with a list of trades and professions, including the housewife, detailed in the hand-written appendix. The report was also concerned with the social class of television viewers and noted that television set ownership was not exclusively for the well-off but that 'if the group is a fair sample, the audience is still predominantly middle class'.[33] However, while such data was collected in the appendix, it was not yet correlated with viewing interests or habits. What is of particular note in the report is the extent to which responses that would, in later years, be understood in terms of gender were here accounted for as representative of the total population. For example, among the main preferences for programme type, the following were listed: 'O.B.'s of Plays (or Variety) from Theatres; News Reels; 'Picture Page'; Light Entertainment (Cabaret, Variety, etc.); O.B.'s of Sporting Events; O.B.'s of other outside events'.[34] In later reports from 1948 to 1951, these were clearly accounted for in terms of male and female preference. However, in the 1936 and 1939 reports, they were classified only in terms of general popularity. Equally, in the 1936 report, viewer requests that 'demonstrations of cooking and fashions should be included in the afternoon and not in the evening programme' were not interpreted in terms of gender.[35] Where gender was discussed it was in terms of the preference for male or female announcers – as in the 1939 report – rather than preferences of male or female viewers.

While it might be assumed that BBC audience research was rudimentary at this time, it is worth noting the extensive use of gender categorization in BBC radio listener research during the same period. In a number of listener reports between 1937 and 1942, there was intense focus on the gender of the audience and the influence it was thought to have on viewing times and preferences. The 1937 Variety Listening Barometer, which surveyed the patterns of listening to various programmes during the day and throughout the week, gathered data on when women listened during the day as well as how this impacted on total listening figures. It also implied that women's listening was inevitably higher than men, noting that 'the disparity is naturally considerable during the daytime'.[36] The 1938 Variety Listening Barometer from the March interim report similarly suggested that 'naturally, since more women are at home than men, afternoon audiences are predominantly feminine'.[37] The Variety Listening Barometer April interim report of the same year noted that the 'size of the feminine audience for daytime programmes is of special interest'.[38] A September 1938 report on Winter Listening Habits produced much more data on the audience share by gender and correlated this with the programme preference and timing of programmes.[39] By 1942, the BBC Listener Research Department had established specific audience panels of around 500 members per panel and aimed at gathering data on audience reactions to specific programmes. Along with two General Listening Panels and a Music Panel, this included a Women's Panel 'mainly for daytime programmes'.[40] Panel members were recruited not through general sampling but from responding to a radio call for volunteer panel members. Silvey himself noted the possibility of volunteer-bias since those volunteering would likely be keener listeners.[41]

However, the researchers were also capable of demonstrating bias in their approach to surveying. A 1942 report on Daytime Programme Repeats, for example, asked the Women's Panel, rather than panels composed of general listeners, about preferences for these daytime programmes.[42] In other words, in assessing the responses to daytime programmes, only women were invited to participate and these were possibly women who already enjoyed such programmes. There are many reasons why there were differences in the extent to which radio listener research had developed where television audience research remained quite sporadic and unsophisticated. Radio broadcasting had been established for some years by the time when audience research formally commenced and had listeners in their millions. It also had a fairly regular schedule of broadcast and a consistent output of programme formats and genres. By comparison, experimental television had a minuscule audience, a fairly sporadic schedule of programmes and, at this stage, an uncertain future. In other words, radio production and broadcast had become efficient and professionalized and, with this, the audience had equally become standardized. The radio audience was, thus, reduced to types and categories, which enabled more large-scale surveying of general trends and habits. This resulted in generalizations about audiences, nowhere more apparent in the often-repeated 'fact' that women watched more during the daytime.

This assertion became common sense and resulted in programme planning that ghettoized women and women's programmes to this period of the day, despite the equally large female audience that listened at other times of the day. Television was yet to shape its audience and its programme schedule in this way. It was concerned with 'everybody' and what this 'everybody' preferred to watch. More rudimentary in its surveys, this was also more egalitarian, although it is possible to see how concerns with social composition were beginning to make their way into surveys. In the early years of television, then, the viewing public was a small but spontaneous, impulsive, changeable group that collectively expressed opinions and attitudes towards the television service and its programmes. Programme-makers were subject to the tastes, interests and behaviour of the audience and (middle-class) women had an equal stake as members of this viewing public. If, for example, a high percentage of the total audience favoured magazine programmes, this was not interpreted through the lens of gender; rather, it was significant in and of itself. As outlined further on, the more gender categorization became a norm within audience measurement, the more this had implications for what types of audiences were valued or not.

This was particularly evident in the post-war years when the television service returned and television audience research began to employ the methods and techniques of categorization established within listener research. The return of BBC television in the post-war years saw the organization more eager for formal audience measurement such as that carried out for radio. The Listener Research Department and the Television Service worked hard to make the case for any form of television audience research. At the same time, the Head of the Television Service and overall coordinator of television programmes, Maurice Gorham, made a number of requests for a means of gathering daily data on the patterns of viewing: a 'television equivalent of the Daily Listening Barometer'.[43] Gorham indicated that he would assist in the generation of a representative panel. Although Gorham and Silvey communicated about the possibility for television audience research, this was initially rejected by the Director General. Gorham's persistence in pursuing this matter suggests the level of urgency felt within the Television Service for the need for an understanding of the television audience. Gorham implored the Director General to change his mind and suggested that it was difficult for the television department to plan productions without knowing what the audience was interested in.[44] In a letter to the Senior Controller, he noted that viewer letters had decreased and asked if he could include a closing announcement in a television programme to solicit feedback from viewers about their opinions on programmes.[45] In a memo to the editor of the *Radio Times*, he went so far as to ask whether it would be possible to have viewers submit their ratings of programmes for publication in the magazine.[46]

Similarly, Silvey began developing plans for television sample panels and worked on preliminary questionnaires for participants.[47] He compiled a draft letter to potential respondents which began 'we want to know as much as possible about our audience'.[48] This knowledge would be gleaned from responses to questions about the make-up of the household as well as their age and gender composition. Unlike the earlier television surveys, newer surveys would concentrate much more on classifying the viewing public. By 1948, the Director General was finally confident enough in the future of television to permit a television audience enquiry and a number of initiatives were undertaken to survey the audience and its interest in programmes.[49] The Listener Research

Department undertook a number of reports on weekly viewing titled the 'Viewers' Vote' scheme as well as reports on the composition and viewing behaviour of the audience such as 'Television: Some Points about the Audience'.[50] The latter demonstrated a far more determined effort to produce a quantifiable and understandable audience where viewing activity could be read in terms of, and in relation to, social categories. Respondents were asked to identify 'how many men, women and children usually watched television when the set was in use'.[51] The use of male and female categories was foregrounded in this report and it is clear to see that the use of gender data was productive of different meanings about and interpretations of the audience.

The gender of the audience was particularly important in the data on viewing times and frequency among the audience. The Frequency of Viewing table, for example, highlighted the differences in numbers of men and women watching at particular times during the day and across the week. The splitting of male and female viewers introduced a new interpretive logic to audience measurement whereby different sense would be made of some members of the audience in comparison to others. This is nowhere more evident in the data that suggested that weekday afternoon viewing was undertaken by 1.4 women in comparison to 0.4 men. In other words, daytime viewing became understood as a predominantly female activity. This resulted in the daytime audience being defined almost exclusively as the 'female audience' despite a significant number of male viewers watching at this time. In addition, while the number of women watching television during the daytime appeared large comparative to men, it was largely in keeping with general trends across the day. The table, thus, gives the impression that women were mainly daytime viewers when, in fact, women watched in fairly consistent numbers throughout the day and week.

The enquiry also noted that the viewer comments suggested an audience that was suburban, middle-class and middle-aged. However, no effort was made to interpret the female audience in relation to these additional variables and, for example, to consider the regional, age or class differences among the total female audience. In other words, the research gave the impression that all women watched a great deal of daytime television when, in fact, the survey sample was representative of those middle-aged, suburban and middle-class women. As Asa Briggs notes, by 1948 television set ownership was distributed among various social classes: 37 per cent of more well-off, 12 per cent of the population in Class 1; 34 per cent of 20 per cent of the middle-income population in Class 2; and 29 per cent of the 69 per cent lower-income population in Class 3.[52] In the BBC enquiry, those middle-class respondents came to represent women of all social classes regardless of the wider demographics of all female viewers. Equally, if the survey respondents were watching during the day, then this was taken as all women, regardless of the proportion of women who might have been elsewhere engaged in non-domestic work and not available to watch daytime television. Essentially, this erased any differences among groups of female viewers and produced a manageable, numerical object called the 'female audience'.

This macro-level view of the audience was quite different to another report published in 1949. The Mass Observation organization – founded in 1937 – undertook regular nationwide studies of the social life of the British public by gathering data through diaries and questionnaires completed by a panel of volunteers.[53] In comparison to the BBC's 1948 enquiry, Mass Observation's Report on Television emphasized the diversity of television viewing amongst women as well as the similarities across social groups. For example, it found that some housewives thought television a waste of time, where others found it a valuable educational resource. The Mass Observation report was based on a survey of 684 people among whom, as Helen Wood notes, only 2 per cent owned a television set.[54] The survey respondents were, according to Wood, largely 'left-leaning and lower middle class because they would have had the time and inclination to commit to the project'.[55] Therefore, they were somewhat similar in class composition to the 1948 BBC enquiry. Despite this, the results of each were different in many respects. Among the female respondents to the Mass Observation survey, many of whom identified themselves as housewives, attitudes towards television were framed in relation to women's identities as productive workers in the home. Some women were concerned that television would be a distraction from other duties and leisure pursuits. Others were concerned that it would confine them to the home and result in less opportunity to be away from the domestic sphere. In this sense, the survey captures what the BBC's audience research did not: the reluctant female audience. The Mass Observation survey's focus on the full and complex spectrum of experiences was in contrast to the reports produced within BBC Listener Research, which imposed order and consistency on audiences and worked to produce knowledge about and meaning from viewer experience.

By the 1950s, the notion that social categories such as age and gender influenced television viewing behaviour had become embedded in the BBC. Audience research continued to define the viewing public by gender and to draw assumptions about viewing patterns based on the gender data gathered. Viewing Panels (formed of samples of households that were representative of the British viewing public) were established and viewers issued with log books which asked them to identify their gender and age alongside their reactions to particular programmes.[56] In his publication 'Methods of Viewer Research Employed by the British Broadcasting Corporation', Silvey maintained that social categories such as gender and age were 'all factors with which programme tastes are liable to be associated'.[57] Although he conceded that there was 'no invariable pattern' and that 'the tastes of men and women are frequently similar and frequently dissimilar', he nonetheless insisted upon the use of these categories in determining some 'basic facts' about the audience.[58] As the 'female audience' became a discursive object within the BBC and this was largely correlated with viewing time, attention to this audience shifted towards daytime viewing. Programme policy increasingly scheduled women's programmes during the daytime rather than in the evening when more women – both in numbers and from different social backgrounds – watched. This was despite Silvey's findings that 'women do not want special women's programmes every afternoon'.[59]

Thus, while audience research was productive of institutional knowledge of the female audience, it continuously had to contend with anomalous and inconsistent behaviour in this group. Research on weekday afternoon viewing, for example, noted that:

> For the various women's programmes an average of 15% of sets are in use but ... there are wide variations. An occasional 'Designed for Women' has touched 29% while one 'Health in the Home' was as low as 5%. The 'viewers per set-in-use' figure is always lower for these programmes, for the obvious reason that they are directed at women who only constitute part of the public (albeit the major part in the afternoon).[60]

In other words, even when the use of the category 'female audience' did little to shed light on viewing patterns and viewer taste, the BBC Audience Research Department continued to deploy it as a meaningful category. This became significant in later years when, as Mary Irwin notes, women's programmes were disappeared from the daytime schedule altogether which, in some ways, demonstrated an undervaluing of women by the BBC.[61]

Ultimately, the 'female audience' was not a social fact. Instead, it was a discursive object used within audience research to make sense of the viewing public. Indeed, during the early years of BBC television, there was no 'female audience' but many female viewers. The creation of the 'audience' enabled the BBC to gain a sense of power in relation to what was once considered a mysterious viewing public. The creation of the 'female audience' allowed the organization to map social inequalities onto programme policy with the result that the female viewer came to occupy as much a marginalized position in relation to the television service as she did in the social sphere. This was done first through the collapsing of vast numbers of culturally, geographically and socio-economically different individuals into one category of 'woman' and, secondly, by developing a programme and schedule strategy that segregated the 'female audience', moving it to daytime schedules that would not interfere with the general audience. Ultimately, the use of gender data within audience research resulted in the ghettoization of female viewers to specific time slots and specific genres. While female viewers, of course, were free to undertake whatever viewing they wished, the institutional production of a 'female audience' meant that women were addressed in gender-specific terms. The production of the 'female audience' within early BBC audience research shows that it was not natural or inevitable that the viewing public would be defined and acted upon according to its gender. Neither was it inevitable that the gender of the viewing public would play so central a role in shaping the programmes and schedules of the television service. However, the methods deployed in measuring the audience resulted in this and an investigation of them reveals the mechanisms by which the viewing public came to be understood as an 'audience'.

Conclusion

The book commenced with the pre-history of television, tracing the ways that television technologies, moving images and entertainment consumption were gendered from the nineteenth century when scientific and popular public imagination began to conceive of the possibility of something like television. In offering a history of the emergence of television, I turned to other media and entertainment that preceded television and which helped to shape what broadcast television became. Television, as I argued, did not create the conditions that resulted in women being its consumers and not producers. Women's place in television culture was influenced by preceding media and entertainment as well as the social and cultural environment in which each emerged. Since early communications and moving image technological innovation was gendered masculine, it was men who helped determine, more generally, what form these innovations would take. While women expressed interest in late nineteenth-century communications technologies, they were excluded from participating in their development. Therefore, women had limited opportunity to shape such technologies, only encountering them after they had been somewhat black-boxed. Since technologies were imagined as masculine, women's engagement with them required some negotiation. This negotiation had begun to take place in other public entertainment forms during the nineteenth century. The theatre and vaudeville, for example, produced the conditions under which women could legitimately, and without concern for reputation, participate in public entertainment.[1] According to Alison Kibler, 'vaudeville ... was a key institution in the transition from a marginalized sphere of popular entertainment, largely associated with vice and masculinity, to a consolidated network of commercial leisure, in which the female consumer was not only welcomed but also pampered'.[2] However, as I argued, women's entry into spaces and venues previously defined as masculine resulted in a good deal of debate and criticism. This is evident in press fascination with the matinee girl who was trivialized and derided for her feminine passions and excessive fandom. Female theatre-goers had, in this sense, low taste. Feminine culture was figured as indicative of low culture.[3] In addition, the opening up of such spaces to women was conditional. In the case of many public entertainments such as theatre, vaudeville, the museum and cinema, women were invited to attend afternoon shows, thus instituting practices of gender segregation. This gender segregation of the audience was intended to provide a safe and comfortable environment for women and children. Yet it also spoke to the anxieties about women's sharing of space as well as moral concerns about hetero-social entertainment experiences.

This paradoxical discourse, in which women were both agents and victims of popular and low culture, was also evident in the case of early moving image exhibitions such as the peep show, the mutoscope and film exhibition. Concern was raised about women's exposure to unsavoury material intended for a male audience. Women were perceived to be at risk of being morally compromised or offended by the salacious and taboo content of early moving image entertainment, particularly the mutoscope. While women were not prevented from using it, dominant public discourse framed the mutoscope as masculine and as a low taste entertainment. However, women were again implicated in the lowering of culture since it was often their image that lured men to the more risqué public entertainments. As I discussed, public entertainment spectacularized the female body such that her body became one of the pleasures offered not only in the mutoscope but in other forms of entertainment and, as Mulvey, Doane and others have pointed out, in cinema.[4] Thus, in the late nineteenth- to early twentieth-century women's relationship to media and public entertainment was in part defined by women's spectacularization on screens and stages since women formed part of the appeal of entertainment. Their relationship with screens was also formed through their consumption of media and entertainment as audiences, purchasers and viewers. Women were discursively produced as the objects of consumption and the consumers, but not the producers of screen media.

Although women were not absent from the sphere of production (after all, those women who appeared on screens were also workers in these industries), they had much less agency and control over it than those men who were the innovators, entrepreneurs and businessmen who formed, shaped and controlled nineteenth- and early twentieth-century communications, media and public entertainment. In addition, despite the large number of women working in these sectors, women's work was often either rendered invisible or treated as feminized and, therefore, less 'productive' and less valued. In accounting for this I assessed the case of the 'television girl' and suggested that she was tasked with embodying a comforting and pleasurable identity that could appropriately represent the promised medium of television. Before television was available on the market, broadcasting organizations wanted to shape the public's ideas about it. In order to generate enthusiasm for it, innovators and business people recruited young, attractive women to be the public face of television. Women such as Hildegarde were featured in newspapers and magazines and spoke on radio programmes about the wonders of the new medium. Where inventors and businessmen worked to convince the public that television was technically and logistically possible, the television girl worked to domesticate it. I argued that women such as Hildegarde and Natalie Towers formed part of the spectacular appeal of television and that their image was used to develop a market for television. In this sense, they also functioned as stand-in commodities for the yet-to-emerge television market. Women's labour both in television communications and as television communicators was, therefore, highly gendered.

This role of women as communicators might have been provisionally accepted in early television work; however, it was a great cause for concern elsewhere. In both the United States and Britain, many women had gained employment as radio announcers, thus having some communicative agency. However, as Michele Hilmes argues, many criticisms were directed at the female radio announcer who was considered unsuitable for the role for a variety of reasons.[5] The reasons offered ranged from the 'fact' that women's voices were not suited to radio broadcasting technology; women's voices were bland and less dynamic; women's voices had no authority in the minds of the listeners (both male and female); and women couldn't legitimately cover male subjects, such as news and sports.[6] In both the United States and Britain, female announcers, who had exercised a good deal of control of radio programme production, found themselves relegated to 'women's interest' broadcasts. It was within this environment that experimental television commenced in the 1930s.

In the case of television, women were considered far more suitable as announcers than was the case with radio. In Britain, the BBC heavily promoted the employment of two women, Jasmine Bligh and Elizabeth Cowell, as its first television announcers. This was accompanied by a number of interviews and reports that emphasized the femininity, grace and class of the two announcers. Many photos of the women appeared in magazines and newspapers and accompanying features wrote of the women's fashion and style. Thus, the women were framed as pleasurable and inoffensive spectacles. In the United States, women such as Betty Goodwin were used to demonstrate the aesthetic and visual potential of television and featured in early demonstrations of programme transmission which were attended and reported on by the press. This difference in the experience of television and radio announcers broadly represents the gendering of audio-visual culture. In the case of radio, Hilmes refers to the anxiety provoked by the disembodied voice which held communicative power.[7] This anxiety perhaps stemmed from the way that the female radio voice deviated so radically from the tradition of objectifying the female form. Because women were more typically represented through images – painting, photography, moving image – her changing role was jarring. As a representation, the viewer retained some authority over woman. In the spectator-representation dynamic, the spectator was afforded some power over the female image.[8] However, in the case of radio, the woman was in a position of authority. She spoke and the listener was spoken to. According to Christine Ehrick, we should think of 'women's radio speech as a performance of the gendered body and as a challenge to the regime wherein women are disproportionately expected to be silent (or at least quiet)'.[9] In contrast, early television representations of women emphasized her appearance rather than her voice and this was more readily accepted in a culture in which the objectification of women was normalized. Indeed, this objectification was part of the means through which women were subjugated and silenced.

Beyond the female announcer, women managed to gain entry into positions in experimental television production that might typically be understood as masculine. As an emerging medium, early television had not yet reproduced the same standardized work processes, hierarchies, male networks and job descriptions. It was, in fact, considered by many in the broadcasting industry a dead-end technology with little future. As such, those who entered this field took on huge risk, particularly if they were moving from radio. In addition, both US and British television were poorly funded. Television production needed to be supported by other financial ventures and businesses. In the United States, experimental stations were not permitted to support advertising until 1941. In Britain, BBC supported experimental television with revenues it received from radio licences. As such, television production remained, during these years, a peripheral activity. This created the conditions under which women could easily flourish. Women, including those who either left or were pushed out of radio, gained work as producers, directors, camera operators, writers and technical crew. The experimental years were, I argued, a period during which women had access to production. I examined the work of women such as Thelma A. Prescott, Frances Buss and Helen Sioussat in the United States and Mary Adams and Mary Allan at the BBC and uncovered evidence of their influence on the development of television formats and genres. Where most histories of women in television had noted women's role in television from the 1950s onwards, we can extend this history further back into the formative and infant years of television. I argued that women's marginalization in television production in the post-war years was a result of a number of factors. The first was that television became a much more 'serious' and professional enterprise following the war. In the United States, commercial television was launched, which in turn provided a source of revenue for television broadcasters. In Britain, this occurred more slowly; yet, the relaunch of the service in 1946, along with an increase in staff, meant that the organization was reluctantly committing to a more professional service. This meant that, following the war, and with television less of a novelty, the place of women in production was not guaranteed to the same extent.

A second factor that influenced women's capacity to work in television was the war itself. In the United States, there was quite a pronounced effect on women's employment as I demonstrated in the case of the Women's Auxiliary Television Technical Staff who were recruited to run the station WBKB in Chicago while male staff were redeployed to aid the war effort. Following the war, the women who had worked for a number of years as camera operators, directors, sound operators and producers were gradually replaced by men who were deemed the natural inheritors of these roles. A third factor was the impact of the gendering of programmes on the gendering of roles and jobs. With the rise of an institutional concept of the female audience in the late 1940s and 1950s, programmes came to be gendered. If daytime women's programmes were thought to appeal to the female audience, then they were also thought to appeal to female producers of television. As production became highly masculinized, women turned to or found themselves working in this genre. By the mid-1950s, both US and British television had become highly gendered in terms of both audience and production. Overall, this sequence of events has left a lasting legacy on television which still bears the trace of such gendering.

Audience research, thus, played a significant role in defining women's relationship to television. As I argued, the audience research that emerged in the broadcast industry drew heavily from the discourse of the female consumer that emerged in late nineteenth- and early twentieth-century market research. In particular, the advertising and consumer industries used market research in order to best position products and services on the market and to make sense of and understand consumer tastes and desires. The rise of consumer societies had, in many ways, been part of a restructuring of social and domestic life. With the increased availability of consumer goods in Western markets, the individual was liberated from many daily tasks. Time was spent not on the production of goods for personal use, but on the consumption of them. Time spent on consumption became interlinked with leisure time, as evident in the rise of department stores and public sites of entertainment. Consumerism afforded one a 'powerful social identity' and sanctioned the expression of desires.[10] The rise of consumer society also resulted in the reorganization of public and private spaces, which made room for sites and products of consumption. This, in turn, was part of a reorganization of gender whereby women's consumption afforded them visibility, agency and power since they were perceived as important actors in the market. As Enrica Asquer notes, 'consumption was … part of a constant renegotiation of gender identities and power inside and outside the domestic sphere'.[11] Although women made up approximately half of the population, they were imagined to hold most household purchasing power.[12] Women were, thus, addressed by advertisers and media as consumers.

Advertising and trade journals perpetuated this notion of the female consumer and gave much attention to proposed strategies that would attract her notice.[13] Market research companies addressed the needs of advertisers by producing knowledge of the female market segment. A host of reports, studies and books emerged that stressed the importance of understanding the motivations, behaviours and lifestyles of women, particularly middle-class, white women since these were of most interest to advertisers. As I argued, these texts did less to account for the actual diversity of women and more to produce a stable, coherent image of the ideal female consumer. Books such as *Selling Mrs. Consumer*[14] painted a picture of the female consumer as eager and responsive to products that helped them in their domestic life and facilitated their role as homemaker. This was an image that said more about the ambitions of advertisers than it did about the diverse lives of ordinary women. When radio broadcasting emerged in the United States in the 1920s, '[broadcasters] turned to the nascent fields of market research and academic radio research' in order to convince advertisers that they could provide them with a valuable market.[15] Within a few years, market research reports of radio consumption had isolated the female listener as a particular object of interest. Since not all female listeners were held as equally valuable in the commercial broadcasting market, researchers honed in on those who were most desirable to broadcasters and advertisers.

These were the middle-class, white women who represented the consumer class. In their efforts to target and isolate this audience, radio researchers turned to the daytime hours where women formed a small but important market. Those women who worked in the household during the day could be specifically addressed through radio and with advertising messages. Ultimately, the housewife became a key demographic of daytime radio and audience researchers produced a multitude of reports to that claimed to be able to map and track the housewife's daytime listening. However, while these reports paid much attention to the housewife's daytime listening habits, they often exaggerated the listening of the housewife and all but ignored other members of the family who may have been listening at this time. By the 1940s, the housewife became synonymous with daytime listening. When television audience research began, a key concern of broadcasters and advertisers was the potential for a daytime audience of housewives. While women may have been able to listen to radio while engaged in domestic work or childcare, broadcasters and advertisers in the United States were not convinced that television could be so accommodating. Television viewing, in their view, required more focused attention. However, broadcasters were driven by a desire to expand television into daytime schedules and then to capitalize on sponsorship and advertising for programmes broadcast during daytime hours. Market researchers and television audience research organizations set about proving the existence of the housewife audience for US daytime television. Despite figures for women's daytime viewing being consistently lower than for their evening viewing, the daytime schedule became the terrain of the housewife. I discussed how this materialized through audience research studies such as those of Videotown and Advertest. Although such audience research was not solely responsible for generating a discourse of the daytime female audience, it contributed to it.

This gendering of television occurred in Britain too, despite the absence of commercial television. The BBC, through its Audience Research Department, set about making sense of the television audience primarily through social composition. Programme preferences, habits and interests were understood as class-specific, age-specific and gender-specific since that was precisely what was measured. Understanding audiences in these terms helped the BBC turn the vast and unstable audience into something far more manageable by using social categories and groupings that the BBC already understood and could make sense of. Programmes were, in turn, addressed to these audiences. In this way, women could be understood as a coherent group that had similar interests and lifestyles and that reflected the ideals and values of broadcasters. Women were, in this sense, homemakers and child carers. Television was addressed to women on these terms. However, as with US television audience research, the female audience produced by the Audience Research Department was more a reflection of its own assumptions about women than it was of the overall female viewing public. To date, very little attention has been paid to how audience research helped to produce the association between the housewife and daytime television and the resulting gendering of television programmes and schedules. Yet the legacy of this persists in contemporary television where audiences are often understood as gendered.

Ultimately, the broadcasting industries that emerged in the United States and Britain were founded upon gender segregation in both television production work and television consumption. The institutional model of gendered television that emerged in the post-war years was, as I argued, shaped by preceding media and entertainment.

By the 1950s, the television industries of each respective nation had actively put into practice the gender segregation of television production and consumption. This has left its trace on television, where women are still largely underrepresented in television work. For example, research has demonstrated that women occupy less than one-third of roles behind the screen on US network television.[16] And in Britain, although there are more women working in the major broadcasters, these numbers reduce as age increases and as with more senior positions.[17] In the area of television consumption, women are still imagined as a subset of the main audience and even new online television platforms have reproduced the gendering of television schedules and genres.[18] While the US and British television industries have become relatively more accessible in recent years, they remain organized according to male working patterns, which women are expected to accommodate themselves to. Understanding how this came to be in the television industry may help us prevent similar patterns emerging in new media industries.

Notes

Introduction

1   Jenkins, Charles F. 1925, *Vision by Radio: Radio Photographs*, Washington, DC: National Capital Press.

2   Hathaway, Kenneth A. 1933, *Television: A Practical Treatise on the Principles upon Which the Development of Television Is Based*, Chicago, IL: American Technical Society.

3   Kerby, Philip. 1939, *The Victory of Television*, New York and London: Harper & Brothers Publishers.

4   Hubbell, Robinson W. 1946, *4,000 Years of Television*, London: G. Harrap & Sons.

5   Jenkins, Charles F. 1931, *The Boyhood of an Inventor*, Washington, DC: National Capital Press.

6   Everson, George. 1949, *The Story of Television: The Life of Philo T. Farnsworth*, New York: W. W. Norton.

7   Russell W. Burns' accounts of television offer a comprehensive technical history of the medium which results in a thoroughly masculine perspective on its development (1986; 1998). Stephen Herbert's three-volume series *A History of Early Television* likewise produces a history of the technical, economic and practical development of television that leaves little space for the inclusion of women (2004a; 2004b; 2004c). Similar television history books that result in male-centred histories of technology, production, economics and business include: George and May Shiers' *Early Television: A Bibliographic Guide to 1940* (1997); *Tube: The Invention of Television* (Fisher and Fisher, 1996); Asa Briggs' five-volume series on British broadcasting, *The History of Broadcasting in the United Kingdom* (1961; 1965; 1970; 1979; 1995); Albert Abramson's two books *The History of Television: 1880–1941* (1987) and *The History of Television: 1942–2000 (2003); Television: The Life Story of a Technology* (Magoun, 2007); and *Stay Tuned: A History of American Broadcasting* (Sterling and Kittross, 2009).

8   Faulkner, Wendy. 2001, 'The Technology Question in Feminism: A View from Feminist Technology Studies.' *Women's Studies International Forum*, 24 (1): 89–90 italics in original.

9   Oldenziel, Ruth. 1999, *Making Technology Masculine: Men, Women and Modern Machines in America, 1870–1945*, Amsterdam: Amsterdam University Press.

10   Wajcman, Judy. 1991, *Feminism Confronts Technology*, Cambridge: Polity Press, 19.

11   Ibid.

12   Ibid., 22.

13   Ibid.

14   Murphy, Kate. 2016, *Behind the Wireless: A History of Early Women at the BBC*, Basingstoke: Palgrave.

15   Hilmes, Michele. 1999, 'Desired and Feared: Women's Voices in Radio History.' In Eds. Mary Beth Haralovich & Lauren Rabinovitz. *Television, History, and American Culture: Feminist Critical Essays*, Durham, NC: Duke University Press.

16   Spigel, Lynn. 1992, *Make Room for TV: Television and the Family Ideal in Postwar America*, Chicago, IL: The University of Chicago Press.

17   Cowan, Ruth Schwartz. 1983, *More Work for Mother: The Ironies of Household Technology from the Open Hearth to the Microwave*, New York: Basic Books.

18   Lohan, Maria & Faulkner, Wendy. 2004, 'Masculinities and Technologies: Some Introductory Remarks.' *Men and Masculinities*, 6 (4): 319.

19   Grazia, Victoria de. 1996, 'Introduction.' In Eds. Victoria de Grazia & Ellen Furlough. *The Sex of Things: Gender and Consumption in Historical Perspective*, Berkeley, CA: University of California Press.

20   Felski, Rita. 1995, *The Gender of Modernity*, Cambridge, MA: Harvard University Press, 62.

21   Grazia, *The Sex of Things*, 7.

22   Scanlon, Jennifer. 2000, 'Introduction.' In Ed. Jennifer Scanlon. *The Gender and Consumer Culture Reader*, New York and London: New York University Press, 7.

23   Ibid.

24   Radway, Janice. 1991, *Reading the Romance: Women, Patriarchy, and Popular Culture*, Chapel Hill and London: University of North Carolina Press.

25 Hansen, Miriam. 2009, *Babel and Babylon: Spectatorship in American Silent Film*, Cambridge, MA: Harvard University Press.

26 Stamp, Shelley. 1998, *Movie-Struck Girls: Women and Motion Picture Culture after the Nickelodeon,* Princeton, NJ: Princeton University Press.

27 Ang, Ien. 1996, *Watching Dallas: Soap Opera and the Melodramatic Imagination*, London and New York: Routledge.

28 See Lynn Spigel's *Make Room for TV: Television and the Family Ideal in Postwar America* (1992); Lynne Joyrich's 1992 article 'All That Heaven Allows: TV Melodrama, Postmodernism, and Consumer Culture.' In Lynn Spigel and Denise Mann's *Private Screening: Television and the Female Consumer* (Minneapolis: University of Minnesota Press); and Patricia Mellencamp's 1992 book *High Anxiety: Catastrophe, Scandal, Age & Comedy*, Bloomington and Indianapolis: Indiana University Press.

29 Huyssen, Andreas, 1986. *After the Great Divide: Modernism, Mass Culture, Postmodernism*, Bloomington, IN: Indiana University Press, 47.

30 Douglas, Ann. 1977, *The Feminization of American Culture*, New York: Alfred A. Knopf.

31 Lysack, Krista. 2008, *Come Buy, Come Buy: Shopping and the Culture of Consumption in Victoria Women's Writing*, Athens, OH: Ohio University Press, 7.

32 See Emma Casey and Lydia Martens' 2012 book *Gender and Consumption: Domestic Cultures and the Commercialisation of Everyday Life* (London and New York: Routledge; Mica Nava's 1996 article 'Modernity's Disavowal: Women, the City and the Department Store.' In Eds. Mica Nava & Alan O'Shea. *Modern Times: Reflections on a Century of English Modernity* (London: Routledge); Shelley Stamp's *Movie-Struck Girls: Women and Motion Picture Culture after the Nickelodeon* (1998, Princeton, NJ: Princeton University Press); Jessica Ellen Sewell's 2011 book *Women and the Everyday City: Public Space in San Francisco, 1890–1915* (Minneapolis, MN: University of Minnesota Press); and Anne Friedberg's 1994 book *Window Shopping: Cinema and the Postmodern* (Berkeley, CA: University of California Press).

33 Friedberg, *Window Shopping: Cinema and the Postmodern*, 36–7.

34 See Donna Halper's 2015 book *Invisible Stars: A Social History of Women in American Broadcasting* (London: Routledge); Cynthia Carter, Gill Branston and Stuart Allan's 1998 book *News, Gender, and Power*

(London: Taylor & Francis); Kate Murphy's 2016 book *Behind the Wireless: A History of Early Women at the BBC* (Basingstoke: Palgrave); Cary O'Dell's 1997 book *Women Pioneers in Television: Biographies of Fifteen Industry Leaders* (Jefferson, NC: McFarland & Company); Linda Seger's 1996 book *When Women Called the Shots: The Developing Power and Influence of Women in Television and Film* (New York: Henry Holt); and Janet Thumim's 2004 book *Inventing Television Culture: Men, Women, and the Box* (Oxford: Oxford University Press) to name but a few books recuperating the histories of women and television.

35 See Marsha F. Cassidy's 2009 book *What Women Watched: Daytime Television in the 1950s* (Austin, TX: University of Texas Press); Lynn Spigel's 1992 book *Make Room for TV: Television and the Family in Ideal Postwar America* (Chicago, IL: University of Chicago Press); Lynn Spigel's 1988 article 'Installing the Television Set: Popular Discourses on Television and Domestic Space, 1948–1955.' *Camera Obscura*, 6 (1); Shaun Moores' 1988 article '"The Box on the Dresser": Memories of Early Radio and Everyday Life.' *Media, Culture & Society*, 10 (1); Susan J. Douglas' 2004 book *Listening In: Radio and the American Imagination* (Minneapolis, MN: University of Minnesota Press); and Mary Beth Haralovich and Lauren Rabinovitz's 1999 edited collection *Television, History, and American Culture: Feminist Critical Essays* (Durham, NC: Duke University Press).

36 Balnaves, Mark, O'Regan, Tom & Goldsmith, Ben. 2011, *Rating the Audience: The Business of Media*, London: Bloomsbury Academic.

37 Ang, Ien. 1991, *Desperately Seeking the Audience*, London: Routledge.

38 Smythe, Dallas. 1977, 'Communications: Blindspot of Western Marxism.' *Canadian Journal of Political and Social Theory*, 1 (3): 1–27.

39 Meehan, Eileen. 2009, 'Gendering the Commodity Audience: Critical Media Research, Feminism, and Political Economy.' In Eds. Meenakshi Gigi Durham & Douglas M. Kellner. *Media and Cultural Studies: Keyworks*. Oxford: Blackwell.

40 Gray, Herman. 1995, *Watching Race: Television and the Struggle for 'Blackness,'* Minneapolis: University of Minnesota Press, 8.

41 Haralovich, Mary Beth. 1989, 'Sitcoms and Suburbs: Positioning the 1950s Homemaker.' *Quarterly Review of Film & Television*, 11 (1): 61–83.

42 Caryl Smith. (1939), *Worcester Democrat and the Ledger-Enterprise*, June: 2, 5.

43 Foucault, Michel. 1994, *The Order of Things: An Archaeology of Human Sciences*, New York: Vintage Books.

44 Fairclough, Norman & Wodak, Ruth. 1997, 'Critical Discourse Analysis.' In Ed. Teun A. van Dijk. *Discourse As Social Interaction*, London: SAGE, 258.

45 Sewell, Philip W. 2014, *Television in the Age of Radio: Modernity, Imagination, and the Making of a Medium*, New Brunswick, NJ: Rutgers University Press.

46 Wodak, Ruth. 1997, *Gender and Discourse*, London: SAGE, 4.

47 Jenkins, *Vision by Radio*.

48 Kerby, *The Victory of Television*.

49 BBC WAC T1/6/1 'Viewers and the Television Service: A Report of an Investigation of Viewers' Opinions in January 1937', 5 February 1937.

50 Advertest. 1949, 'Study of Daytime Television.' *The Television Audience of Today*, 1 (1). NBC Files: Box 193, folder 2. Wisconsin Historical Society Archives.

51 Lind, Rebecca Ann. 2006, 'Understanding the Historical Context of Race and Gender in Electronic Media.' In Ed. Donald G. Godfrey. *Methods of Historical Analysis in Electronic Media*, London: Lawrence Erlbaum Publishing, 246.

52 Hodgson, Violet. 1931, 'Television Goes to Sea.' *Radio News*, November 1931, 386–8.

53 Frederick, Christine McGaffey. 1929, *Selling Mrs. Consumer*, New York: The Business Bourse.

Chapter 1

1 Kitch, Caroline. 2009, *The Girl on the Magazine Cover: The Origins of Visual Stereotypes in American Mass Media*, Chapel Hill, NC: University of North Carolina Press.

2 Cott, Nancy. 1987, *The Grounding of Modern Feminism*, New Haven, CN: Yale University Press.

3 Gleadle, Kathryn. 2017, *British Women in the Nineteenth Century*, Basingstoke: Palgrave, 5.

4 Ware, Susan. 1993, *Still Missing: Amelia Earhart and the Search for Modern Feminism*, New York: W. W. Norton.

5   Oakley, Ann. 1974, *Woman's Work: The Housewife Past and Present*, New York: Vintage Books.

6   Cott, *The Grounding of Modern Feminism*.

7   Marcellus, Jane. 2011, *Business Girls & Two-Job Wives: Emerging Media Stereotypes of Employed Women*, Cresskill, NJ: Hampton Press.

8   Marks, Patricia. 2014, *Bicycles, Bangs, and Bloomers: The New Woman in the Popular Press*, Lexington, KY: The University Press of Kentucky.

9   Petro, Patrice. 1986, 'Mass Culture and the Feminine: The "Place" of Television in Film Studies.' *Cinema Journal*, 25 (3): 5–21.

10   Paul Nipkow discovered in the late 1880s scanning which was crucial for transmission; Karl Ferdinand Braun invented the cathode ray tube in the late 1800s; John Logie Baird invented the system for mechanical television in the 1920s; Philo T. Farnsworth developed the first electronic television in the late 1920s.

11   See 'A Female Inventor' (1890), *Los Angeles Herald*, November: 24, 7; 'Women as Inventors' (1893), *New York Tribune*, February: 2, 5.

12   Stanley, Autumn. 1995, *Mothers and Daughters of Invention: Notes for a Revised History of Technology*, New Brunswick, NJ: Rutgers University Press.

13   Oldenziel, *Making Technology Masculine: Men, Women and Modern Machines in America, 1870–1945*, 10.

14   Wajcman, Judy. 2013, 'Addressing Technological Change: A Challenge to Social Theory.' In Eds. Morton Winston & Ralph Edelbach. *Society, Ethics, and Technology*, Fifth Edition. Boston, MA: Cengage Learning, 100.

15   Cockburn, Cynthia. 1985, *Machinery of Dominance: Women, Men and Technical Know-How*, London: Pluto Press.

16   Wajcman, *Feminism Confronts Technology*.

17   Cockburn, Cynthia. 1992, 'The Circuit of Technology: Gender, Identity and Power.' In Eds. Roger Silverstone & Eric Hirsch. *Consuming Technologies: Media and Information in Domestic Spaces*, London: Routledge, 38.

18   Paiva, Adriano de. 1880, 'La téléphonie, la télégraphie et la télescopie.' In *O Instituto* [online]. Available at: http://webcache.googleusercontent.com/search?q=cache:hMaY3AH5tbUJ:histv2.free.fr/de_paiva/telescopie1.htm+&cd=11&hl=en&ct=clnk&gl=ie [Accessed on 12 December 2017].

19   Paiva, 'La téléphonie, la télégraphie et la télescopie.' Italics added.

20 The implications of the association of women with nature and man with culture and science have been discussed in Simone de Beauvoir's *The Second Sex* (2009/1949), in Sherry Ortner's 'Is Female to Male as Nature Is to Culture?' (1972) and Ludmilla Jordanova's *Sexual Visions: Images of Science and Medicine between the Eighteenth and Twentieth Centuries* (1989). Each of these authors considers the means by which the social produced and practised beliefs of gender situate men as masters, agents and cultural beings and women as irrational, primitive and associated with nature.

21 Hunt, Verity. 2014, 'Electric Leisure: Late Nineteenth-Century Dreams of a Remote Viewing by "Telectroscope".' *Journal of Literature and Science*, 7 (1): 55–76.

22 George du Maurier, 'Edison's Telephonoscope,' *Punch's Almanack* for 1879, 9 December 1878, 52.

23 Shiers, George & Shiers, May. 1997, *Early Television: A Bibliographic Guide to 1940*, New York and London: Garland Publishing, 15.

24 Robida, Albert. 1883/2004, *The Twentieth Century*. Trans. Philippe Willems. Ed. Arthur B. Evans. Middletown, CT: Wesleyan University Press.

25 Herbert, Stephen. 2004, *Eadweard Muybridge: The Kingston Museum Bequest*, East Sussex: The Projection Box, 115.

26 Musser, Charles. 1994, *The Emergence of Cinema: The American Screen to 1907*, Volume 1. Berkeley, CA: University of California Press.

27 In fact, the founder of the American Mutoscope Company, William K.L. Dickson had worked on the kinetoscope at the Edison Manufacturing Company and was responsible for many of the inventions that came out of the company. Dickson left Edison to form his own rival company.

28 Musser, *The Emergence of Cinema*, 75.

29 Ibid., 78.

30 Musser, Charles. 2004, 'At the Beginning: Motion Picture Production, Representation and Ideology at the Edison and Lumiére Companies.' In Eds. Lee Grieveson & Peter Krämer. *The Silent Cinema Reader*, London: Routledge, 22.

31 Ibid., 23.

32 'Refined Entertainment' (1903), *The Wichita Daily Eagle*, June: 11, 6.

33 For example, in Canada, a number of petitions were brought before the House of Commons by the Women's Temperance Union and for the

prohibition of 'exhibitions of an immoral character, by kinetoscope, &c'. See Parliament. House of Commons of the Dominion of Canada. *Petitions*, 17 May 1897, p. 157 and Parliament. House of Commons of the Dominion of Canada. *Petitions*, 2 June 1897, 205. By 1907, Chicago introduced censorship by prohibiting 'the exhibition of obscene and immoral pictures and regulating the exhibition of pictures of the classes and kinds commonly shown in mutoscopes, kinetoscopes, cinematographs, and penny arcades'. All Amendments to The Revised Municipal Code of Chicago of 1905 (passed 20 March 1905): And All New General Ordinances and Ordinances Creating Prohibition and Local Option Districts Passed ... from 20 March 1905 to 1 January 1909. Supplement number III. 1907. Ordinance: *Mutoscopes, Etc., Regulating the Exhibition of Pictures in*, 4 November 1907, 335.

34 'Phonograph Men's Lesson in Morality' (1899), *The San Francisco Call*, April: 21, 6.

35 For example, *The Birth of the Pearl* (American Mutoscope Company, 1903) or *Dorolita's Passion Dance* (1894) and which was withdrawn from exhibition.

36 'Parisian Topics: The Exhibition Closing' (1900), *The Standard* (UK), August: 24, 5.

37 'An Absent-Minded Thief' (1900), *The Illustrated Police News*, December: 8, 2.

38 'Ladies' Column' (1901), *Evening Telegraph*, June: 12, 6.

39 Ibid., 6.

40 See Parrett, Catriona. 2001, *More Than Mere Amusement: Working-Class Women's Leisure in England, 1750–1914*, Boston, MA: Northeastern University Press; Peiss, Kathy. 2011, *Cheap Amusements: Women and Leisure in Turn-of-the-Century New York*, Philadelphia, PA: Temple University Press; and Rappaport, Erika Diane. 2000, *Shopping for Pleasure: Women in the Making of London's West End*, Princeton, NJ: Princeton University Press.

41 See Scott, Anne Firor. 1992, *Natural Allies: Women's Associations in American History*, Urbana and Chicago: University of Illinois Press; Doughan, David & Gordon, Peter. 2007, *Women, Clubs and Associations in Britain*, London: Routledge.

42 See Rappaport, Erika Diane, *Shopping for Pleasure: Women in the Making of London's West End*; Schweitzer, Marlis. 2009, *When Broadway Was*

the Runway: Theatre, Fashion, and American Culture*, Philadelphia, PA: University of Pennsylvania Press.

43 Bates, A.W. 2008, '"Indecent and Demoralising Representations": Public Anatomy Museums in mid-Victorian England.' *Journal of Medical History*, 52 (1): 1–22.

44 'Barnum's Museum, Corner Chestnut and Seventh Streets, Philadelphia' (1849), *Lancaster Examiner and Herald*, July: 11, 1.

45 Pearson, Susan T. 2008, '"Infantile Specimens": Showing Babies in Nineteenth-Century America.' *Journal of Social History*, 42 (2): 241–370.

46 'Baby Shows, Godey's Lady's Book. December 1855.' 2018. The Lost Museum Archive. Available at: https://lostmuseum.cuny.edu/archive/baby-shows-godeys-ladys-book-december. [Accessed on 26 July 2018].

47 Pandora, Katherine. 2017, 'The Permissive Precincts of Barnum's and Goodrich's Museums of Miscellaneity: Lessons in Knowing Nature for New Learners.' In Eds. Carin Berkowitz & Bernard Lightman. *Science Museums in Transition: Cultures of Display in Nineteenth-Century Britain and America*, Pittsburgh, PA: University of Pittsburgh Press.

48 *Punch or the London Charivari 19*. 1850, 'Jenny Lind and the Americans' – Coronation of Jenny the First – Queen of the Americans.' Cartoon 5 October 1850, 146.

49 Staats, Eleanor E. 1980, 'Some Don'ts for Girls.' *The Ladies' Home Journal*, January 1890, 10.

50 Ryan, Mary P. 1981, *Cradle of the Middle-class: The Family in Oneida County, New York, 1790–1865*, Cambridge: Cambridge University Press.

51 'KumfurtShoe & Co. Big Shoe Sale' (1897), *The Sun*, March: 28, 12.

52 Giles, Judy. 2007, 'Class, Gender and Domestic Consumption in Britain 1920–1950.' In Eds. Emma Casey & Lydia Martens. *Gender and Consumption: Domestic Cultures and the Commercialisation of Everyday Life*, London: Routledge.

53 Huyssen, Andreas. 1986, *After the Great Divide: Modernism, Mass Culture, Postmodernism*, Bloomington, IN: Indiana University Press, 47.

54 'Free Lectures to Ladies Only' (1890), *The Salt Lake Herald*, June: 25, 8.

55 'A Model Café: How the Brunswick Tables and Serves Its Guests' (1882), *Wheeling Sunday Register*, September: 2, 1.

56 'Danal's Staking Rink' (1886), *The Daily Telegraph*, February: 1, 3.

57 Lysack, *Come Buy, Come Buy*.

58 Arceneaux, Ronald J. 2007, 'Noah'. *Department Stores and the Origins of American Radio Broadcasting, 1910–1931*, PhD Dissertation. University of Georgia. Available at: https://getd.libs.uga.edu/pdfs/arceneaux_ronald_j_200705_phd.pdf. [Accessed on 26 July 2020].

59 McCarthy, Anna. 2001, *Ambient Television: Visual Culture and Public Space*, Durham, NC: Duke University Press, 65.

60 Rappaport, *Shopping for Pleasure*.

61 Lysack, *Come Buy, Come Buy*, 46.

62 'Desperate Bargains' (1876), *Belfast Telegraph*, August: 14, 4.

63 Sewell, *Women and the Everyday City*.

64 'The Eden Musee: Lecture Room, Curiosity Hall and Bijou Theatorium' (1877), *The Omaha Daily Bee*, December: 18, 3.

65 'Utterly Unprecedented Success of the Great Adelphi Company' (1874), *Chicago Daily Tribune*, February: 5, 3.

66 Gardner, Viv. 2015, 'The Theatre of the Flappers?: Gender, Spectatorship and the "Womanisation" of Theatre 1914–1918.' In Ed. Andrew Maunder. *British Theatre and the Great War, 1914–1919*, London: Palgrave Macmillan.

67 Butsch, Richard, *The Making of American Audiences: From Stage to Television, 1750–1990*, Cambridge: Cambridge University Press, 124–5.

68 'Notes by the Matinee Man' (1907), *Los Angeles Herald*, March: 17, 12.

69 'Matinee Girls between Acts' (1896), *The Morning Times*, March: 8, 21.

70 Butsch, *The Making of American Audiences: From Stage to Television, 1750–1990*, 76.

71 'The Matinee Girl and Her Footlights Hero' (1903), *The Washington Times*, May: 24, 6.

72 Ibid, 6.

73 'A New Matinee Girl' (1904), *The Salt Lake Tribune*, April: 3, 6.

74 Ibid., 6.

75 Erdman, Andrew L. 2007, *Blue Vaudeville: Sex, Morals and the Mass Marketing of Amusement, 1895–1915*, London: McFarland & Company.

76 Kitch, *The Girl on the Magazine Cover*, 6–8.

77 Agnew, Jeremy. 2011, *Entertainment in the Old West: Theater*, Music, Circuses, Medicine *Shows, Prizefighting and Other Popular Amusements*, Jefferson, NC: McFarland & Co, 59.

78 Allen, Robert Clyde. 2000, *Horrible Prettiness: Burlesque and American Culture*, Chapel Hill: University of North Carolina Press.

79 Butsch, *The Making of American Audiences: From Stage to Television, 1750–1990*, 94.

80 Ibid., 96.

81 Wheeler, Leigh Ann. 2004, *Against Obscenity: Reform and the Politics of Womanhood in America, 1873–1935*, Baltimore, MD: Johns Hopkins University Press.

82 Welter, Barbara. 1976, *Dimity Convictions: The American Woman in the Nineteenth Century*, Athens: Ohio University Press.

83 Mizejewski, Linda. 1999, *Ziegfeld Girl: Image and Icon in Culture and Cinema*, London: Duke University Press, 68.

84 Grazia, *The Sex of Things*, 280.

85 Mulvey, Laura. 1975, 'Visual Pleasure and Narrative Cinema.' *Screen*, 16 (3): 6–18.

Chapter 2

1 John Logie Baird demonstrated his mechanical television in Selfridges department store in London in March 1925. This was a demonstration of a televised image of a puppet since it was not yet possible to transmit images of people. In 1926, he held another television demonstration of a human performer – a woman – that is often referred to as the first television demonstration.

2 Ernst Alexanderson demonstrated television transmission in January 1928. He believed that television would initially be transmitted to theatres before then being used in the home. (Edison Tech Center, 2014).

3 Radio Corporation of America (RCA) demonstrated television in the 'Hall of Televisions' at the New York World's Fair in 1939. RCA used the World's Fair to launch regular broadcast programmes as well as to advertise its television sets and to pitch television as the next big consumer product to follow radio. See von Schilling, James A. 2013, *The Magic Window: American Television 1939–1953*. New York: Haworth Press.

4   See Sewell's discussion of a 1927 cover of *Radio News* magazine which
    suggests women's misuse of television. Sewell, *Television in the Age of Radio*.

5   Boddy, William. 1993, *Fifties Television: The Industry and Its Critics*,
    Chicago, IL: University of Illinois Press.

6   In 1927, Bell Laboratories demonstrated what was effectively a video call
    of Herbert Hoover in Washington speaking with those in the laboratory
    in New York. See 'Far-Off Speakers Seen as Well as Heard Here in a Test
    of Television' (1927), *The New York Times*, April: 8, 20.

7   'How Will Television Affect the Theater?' (1931), *The Sunday Star*,
    January: 4, 5.

8   In addition to existing radio broadcasters, licences were issued by the
    FCC in the United States and the Postmaster General in Britain to
    newer entries to the television sector, including General Electric Co.,
    entertainment corporations such as Balaban & Katz and Paramount, and
    newly formed television companies such as Charles Jenkins Laboratories.

9   Burns, Russell W. 1998, *Television: An International History of the
    Formative Years*, London: The Institution of Electronic Engineers.

10 Wheatley, Helen. 2016, 'Television in the Ideal Home.' In Eds. Rachel
    Moseley, Helen Wheatley & Helen Wood. *Television for Women: New
    Directions*, London: Routledge.

11 Boddy, *Fifties Television: The Industry and Its Critics*, 28–9.

12 Many newspapers and magazines featured images of men operating
    television. Women would occasionally feature as the sitting subjects
    that were televised or as performers. See 'Seeing as You Hear Over the
    Radio' (1929), *The Evening Star*, January: 16, 13; 'Radio Men Witness
    Demonstration of Television' (1928), *The Evening Star*, May: 5, 2;
    'Television's New Aerial "Eye" for War and Peace Broadcasting' (1929),
    New Britain Herald, August: 23, 25.

13 Arceneaux, Noah. 2018, 'Department Stores and Television.' *Journalism
    History*, 43 (4): 219–27.

14 As Mark Aldridge notes, it was quite common for Selfridges to hold
    demonstrations and host exhibitions that enhanced its reputation as a
    lavish department store that was aligned with social and technological
    progress. See Aldridge, Mark. 2011, *The Birth of British Television: A
    History*, London: Palgrave Macmillan.

15 Burns, R.W. 2000, *John Logie Baird: Television Pioneer*, London: The Institution of Electronic Engineers. According to Burns, Baird was less interested in the publicity than he was in the pay he would receive for holding the demonstrations. It was Selfridges that wished to exhibit television. Baird, in fact, was quite dismissive of the shoppers that came to view the apparatus.

16 Novotny, Patrick. 2014, *The Press in American* Politics, *1787–2012*, Santa Barbara, CA: Praeger, 105.

17 Jenkins, *Vision by Radio*, 12.

18 Ibid., 12.

19 Ibid., 13.

20 Arceneaux, 'Department Stores and Television.'

21 Becker, Ron. 2001, 'Hear-and-See Radio' in the World of Tomorrow: RCA and the Presentation of Television at the World's Fair, 1939–1940.' *Historical Journal of Film, Radio and Television*, 21 (4): 361–78.

22 Novotny, *The Press in American Politics, 1787–2012*, 106.

23 Burns, *Television: An International History of the Formative Years*.

24 Dinsdale, A. 'A Successful Public Demonstration of Television between Washington and New York' (1927), *The Wireless World*, June: 1, 681.

25 Ibid., 1.

26 Rappoport, Leon. 2005, *Punchlines: The Case for Racial, Ethnic, and Gender Humor*, London: Praeger, 103.

27 Ibid., 109.

28 *The New York Times*, 'Far-Off Speakers Seen as well as Heard Here in a Test of Television,' 20.

29 Mercer, David. 2006, *The Telephone: The Life Story of a Technology*, Westport, CT: Greenwood Press, 51–2.

30 Green, Venus. 2001, *Race on the Line: Gender, Labor, and Technology in the Bell System, 1880–1980*, Durham, NC: Duke University Press.

31 This cultivation of the television switchboard operator role as nurturing and supportive would also be evident in the press response to the BBC's Joan Miller whose role on BBC programme *Picture Page* was to 'connect' the audience to the guests on the programme.

32 Hilmes, Michele. 1997, *Radio Voices: American Broadcasting, 1922–1952*, Minneapolis, MN: University of Minnesota Press, 136.

33 McLean, Donald F. 2000, *Restoring Baird's Image*, London: The Institute of Electrical Engineers, 175.

34 Burns, *Television: An International History of the Formative Years*, 287.

35 *The Man with the Flower in His Mouth* (1930), [Television Programme] BBC.

36 Conway, Mike. 2009, *The Origins of Television News in American: The Visualizers of CBS in the 1940s*, New York: Peter Lang, 16.

37 Douglas, Susan J. 1999, *Listening in: Radio and the American Imagination*, Minneapolis: University of Minnesota Press.

38 Murphy, Kate. 2016, '"New and Important Careers": How Women Excelled at the BBC, 1923–1939.' *Media International Australia*, 161 (1): 18–27.

39 Crook, David. 2007, 'School Broadcasting in the United Kingdom: An Exploratory History.' *Journal of Educational Administration and History*, 39 (3): 217–26.

40 Murphy, Kate. 2019, 'Relay Women: Isa Benzie, Janet Quigley and the BBC's Foreign Department, 1930–38.' *Feminist Media Histories*, 5 (3): 114–39.

41 Jones, Allan. 2011, 'Mary Adams and the Producer's Role in Early BBC Science Broadcasts.' *Public Understanding of Science*, 21 (8): 968–83.

42 Halper, Donna. 2008, 'Speaking for Themselves: How Radio Brought Women into the Public Sphere.' In Ed. Michael C. Keith. *Radio Cultures: The Sound Medium in American Life*. New York: Peter Lang.

43 Halper, *Invisible Stars: A Social History of Women in American Broadcasting*, 32.

44 Keeler, Amanda. 2017, '"A Certain Stigma" of Educational Radio: Judith Waller and "Public Service" Broadcasting.' *Critical Studies in Media Communication*, 34 (5): 495–508.

45 O'Dell, *Women Pioneers in Television: Biographies of Fifteen Industry Leaders*.

46 Ware, Susan. 2005, *It's One O'Clock and Here Is Mary Margaret McBride: A Radio Biography*, New York: New York University Press.

47 Hilmes, *Radio Voices*, 141.

48 Murphy, *Behind the Wireless*.

49 Murray, Susan. 2005, *Hitch Your Antenna to the Stars: Early Television and Broadcast Stardom*, London: Routledge, XIII.

50 'Television Is Here' (1931), *Radio Digest*, September: 17, 1.

51 Ibid., 2.

52 Ibid., 39.

53 Ibid., 33–5.

54 Murray, *Hitch Your Antenna to the Stars*.

55 'Mechanical Television: The Queen's Messenger' 1999, *Early Television Museum* [online]. Available at: http://www.earlytelevision.org/queens_messenger.html [Accessed on 12 December 2017].

56 Ibid.

57 'Baby-faced Types Bad for Television' (1931), *Evening Star*, May: 31, 9.

58 Ibid.

59 'Pretty Girls' (1929), *The Bismark Tribune*, July: 6, 4.

60 'Back to the Kilts Again' (1928), *The Derby Daily Telegraph*, October: 1, 6.

61 'Sent Out Backwards' (1929), *Yorkshire Evening Post*, September: 30, 7.

62 'Television Girl' (1930), *Hartlepool Northern Daily Mail*, October: 2, 4.

63 Davis, D. 'Television Peeps around the Corner' (1931), *What's on the Air*, June, 3.

64 'Girl Dummy Aids Television Tests' (1937), *Evening Star*, September: 12, B-3.

65 Davis, 'Television Peeps around the Corner,' 3.

66 *Mid-Week Pictorial*. 1931, 'Miss Television.' No date, front cover illustration. Available at: http://genealogyimagesofhistory.com/images3/Natalie-Towers.jpg [Accessed on 7 June 2019].

67 Babst, Jacob L. & Tribe, Ivan M. 2019, *Beryl Halley: The Life and Follies of a Ziegfeld Beauty, 1897–1988*, Jefferson, NC: McFarland, 165.

68 Dunlap, Orrin E. 1932, *The Outlook for Television*, New York & London: Harper & Brothers Publishers, 227–8.

69 Ibid.

70 'Television's Eye Often Deceiving' (1931), *The Daily Illini*, October: 28, 4.

71 Dyer, Richard. 2004, *Heavenly Bodies: Film Stars and Society*, Second Edition. London: Routledge, 5.

72 Watkins Reeves, Mary. 1937, 'Who Are the First Real Stars of Television?' *Radio Mirror*, March 1937, 47.

73 Ibid.

74 'First "Television Tea" Held by Women's Club' (1931), *The Watchman*, March: 5, 5.

Chapter 3

1   Murphy, *Behind the Wireless*.
2   Logan, Anne. 2020, 'Gender, Radio Broadcasting and the Role of
    the Public Intellectual: The BBC Career of Margery Fry, 1928–1958.'
    Historical Journal of Film, Radio and Television, 40 (2): 389–406.
3   *Women's Hour* (1923–5), [Radio Programme] BBC.
4   *Household Talks* (1927–9), [Radio Programme] BBC.
5   *The Week in Parliament/The Week in Westminster* (1929–), [Radio
    Programme] BBC.
6   Murphy, *Behind the Wireless*.
7   Ibid., 6.
8   Ibid., 122.
9   Irwin, Mary. 2011, 'What Women Want on Television: Doreen Stephens
    and BBC Television Programmes for Women, 1953–1964.' *Westminster
    Papers in Communication and Culture*, 8 (3): 99–122.
10 See Newcomb, Horace. 2014, *Encyclopedia of Television*, Second Edition.
    London and New York: Routledge; Inglis, Ruth. 2003, *The Window in
    the Corner: A Half-Century of Children's Television*, London: Peter Owen;
    Holmes, Su. 2016, 'Revisiting Play School: A Historical Case Study of
    the BBC's Address to the Pre-school Audience.' *The Journal of Popular
    Television*, 4 (1): 29–47.
11 Sandon, Emma. 2018, 'Engineering Difference: Women's Accounts of
    Working as Technical Assistants in the BBC Television Service between
    1946 and 1955.' *Feminist Media Histories*, 4 (4): 8–32.
12 Terkanian, Kathryn. 2019, *Women, Work, and the BBC: How Wartime
    Restrictions and Recruitment Woes Reshaped the Corporation, 1939–45*,
    PhD Thesis. Bournemouth University.
13 Burns, *John Logie Baird*, 154.
14 Ibid., 166.
15 Burns, Russell W. 1986, *British Television: The Formative Years*, London:
    The Institution of Electronic Engineers, 283.
16 Hendy, David. 2019, 'The BBC Steps in: 1929–1935.' Available at: https://
    www.bbc.com/historyofthebbc/100-voices/birth-of-tv/the-bbc-steps-in
    [Accessed on 1 October 2019].
17 Ibid.

18 Baily, Kenneth, 'Here's Looking at You' (n.d.), Available at: https://
    www.teletronic.co.uk/pages/history_of_the_bbc_3.html [Accessed on 8
    September 2019].

19 Ibid.

20 'Television Announcer. Search for a "Superwoman"' (1935), *Sydney
    Morning Herald*, November: 14, 11.

21 Murphy, *Behind the Wireless*, 241–4.

22 Norman, Bruce. 1984, *Here's Looking at You: The Story of British
    Television, 1908–1939*, London: Royal Television Society, 122–3.

23 'The Position with Regard to Television' (1936), *Gloucestershire Echo*,
    January: 6, 4.

24 'Television Announcers' (1935), *Lancashire Evening Post*, December: 14, 4.

25 Norman, *Here's Looking at You*, 123.

26 'Television Face' (1935), *Yorkshire Post and Leeds Intelligencer*, November:
    14, 8.

27 Ibid.

28 Ibid.

29 Ibid.

30 'B.B.C. to Have Women Announcers?' (1935), *Nottingham Evening Post*,
    July: 2, 6.

31 'The Woman Announcer' (1935), *Television and Short-Wave World*
    December, 694.

32 Murphy, *Behind the Wireless*, 19.

33 Bennett, James. 2010, *Television Personalities: Stardom and the Small
    Screen*, London: Routledge.

34 Baily quoted in Bennett, *Television Personalities*, 77.

35 Fletcher, G. 'People You See: The Television Announcers' (1937), *Radio
    Times*, January: 8, 8–9.

36 'Women Announcers' (1934), *Nottingham Journal*, April: 10, 5.

37 Zakharine, Dmitri. 2013, 'Voice- E-Voice-Design- E-Voice-Community:
    Early Public Debates about the Emotional Quality of Radio and TV
    Announcers' Voices in Germany, the Soviet Union and the USA (1920–
    1940).' In Eds. Dmitri Zakharine & Nils Meise. *Electrified Voices: Medial,
    Socio-Historical and Cultural Aspects of Voice Transfer*, Gottingen: V & R,
    214.

38 'Our London Letter' (1936), *Northern Whig*, May: 14, 6.

39 'The Television Announcer-Hostesses' (1936), *Television and Short-Wave World*, May, 324.

40 Bennett, *Television Personalities*, 77–8.

41 'Radio Notes' (1936), *Birmingham Daily Gazette*, September: 1, 8.

42 'B.B.C. Announcers' (1935), *Yorkshire Post and Leeds Intelligencer*, May: 30, 7.

43 BBC WAC Staff File Jasmine Bligh Memo from Gerald Cock about Jasmine Bligh's contract, 1 February 1938.

44 BBC WAC Staff File Memo from Gerald Cock to Jasmine Bligh, 12 December 1938.

45 BBC WAC Staff files Jasmine Bligh. R.A. Rendall record of interview with Jasmine Bligh, 24 April 1939.

46 Murphy, Catherine. 2011, *'On an Equal Footing with Men?' Women and Work at the BBC, 1922–1939*, PhD Thesis. Goldsmiths College: University of London, 21.

47 Ibid.

48 *A Laundry Demonstration* (1936), [Television Programme] BBC.

49 'Programmes: Women's Interest. A Laundry Demonstration' (1936), *Radio Times*, November: 13, 93.

50 BBC WAC T1/6/1 'Viewers and the Television Service: A Report of an Investigation of Viewers' Opinions in January 1937', 5 February 1937.

51 Adams, Mary (Producer). (12 January 1937), *World of Women: Setting a Play* [Television Broadcast]. London: BBC.

52 McArthur, M. 'The World of Women: "Setting a Play"' (1937), *Radio Times Television Supplement*, January: 8, 6.

53 Evans, Myfanwy (Producer). (22 July 1937), *Expedition on a Bicycle* [Television Broadcast]. London: BBC.

54 Adams, Mary (Producer). (8 February 1937), *World of Women: Painting and Pottery* [Television Broadcast]. London: BBC.

55 Dame Laura Knight, D.B.E, R.A., L.L.D. 'The World of Women: Painting and Pottery' (1937), *Radio Times Television Supplement*, February: 5, 5.

56 Clark, D. 'World of Women: "New Series 1. Sculpture in Wood, Bronze, and Stone"' (1937), *Radio Times Television Supplement*, April: 9, 2.

57 Ibid.

58 Field, M. 'The World of Women: "The Making of Documentary and "Secret of Nature" Films"' (1937), *Radio Times Television Supplement*, January: 22, 6.

59 'Filming through a Microscope' (1937), *Radio Times Television Supplement*, January: 22, 5.

60 Ibid.

61 Murphy, *Behind the Wireless*.

62 Murphy, *On an Equal Footing with Men*, 21.

63 Ibid.

64 Terkanian, *Women, Work, and the BBC*, 75–6.

65 Murphy, *On an Equal Footing with Men*, 27.

66 Currie, Tony. 2004, *British Television 1930–2000*, Second Edition. Devon: Kelly Publications; Cornell, Paul, Day, Martin & Topping, Keith. 2004, *The Classic British Telefantasy Guide*, London: Hachette UK.

67 Terkanian, *Women, Work, and the BBC*, 74.

68 *Picture Page* (1936–9; 1946–52), [Television Programme] BBC.

69 *Julius Caesar* (24 July 1938), [Television Broadcast] Producer Dallas Bowers, London: BBC.

70 *R.U.R.* (11 February 1938), [Television Broadcast] Producer Jan Bussell, London: BBC.

71 Allan, M. 'Make-Up' (1937), *Radio Times Television Supplement*, April: 2, 4–5.

72 'Here's Looking at Them!' (1936), *Radio Times Television Supplement*, October: 23, 10–11.

73 'Television Make-Up' (1936), *Wireless World*, August: 14, 161.

74 'High Speed Make-Up' (1936), *Wireless World*, October: 16, 405.

75 Hunt, K.P. 1937, 'A Studio Screen: A Monthly Causerie on Television Personalities and Topics.' *Television and Short-Wave World*, February 1937, 97.

76 'A Forecast of Transmission Improvements' (April 1937), *Television and Short-Wave World*, 226.

77 'Spends Twelve Hours a Day on Make-Up' (1937), *The Sunday Times* (Perth, Australia), November: 7, 36.

78 Baily, Kenneth. 1950/2001, 'A Personal Account of Television's Early Days.' Reprinted in *Teletronic: The Original Television History Site*. Available at: http://www.teletronic.co.uk/herestv4.htm [Accessed on 12 December 2017].

79 Bailey, Michael. 2009, 'The Angel in the Ether: Early Radio and the Construction of the Household.' In Ed. Michael Bailey. *Narrating Media History*, London: Routledge.

80 Murphy, *'On an Equal Footing with Men?'* 170.

81 *Problems of Heredity* (1928), [Radio Programme] BBC.

82 Ibid.

83 *Clothes-line* (1937), [Television Programme] BBC.

84 *Friends from the Zoo* (1937), [Television Programme] BBC.

85 *Architecture* (1937), [Television Programme] BBC.

86 *The Future of Television* (1937), [Television Programme] BBC.

87 *Clothes through the Centuries* (1938), [Television Programme] BBC.

88 *Artists and Their Work* (1938), [Television Programme] BBC.

89 *Spelling Bee* (1938), [Television Programme] BBC.

90 *Sight and Sound* (1939), [Television Programme] BBC.

91 More O'Ferrall was variety and drama producer and director in the experimental years of television. He produced the variety programme *Picture Page.*

92 Cecil Lewis was a veteran of BBC radio having joined the company in 1922. He moved to television working on Outside Broadcasts and Talks from 1936 to 1937. Adams replaced him when he left the BBC.

93 Cecil Madden moved from radio into television in 1936 and was programme planner and producer of *Picture Page* and other programmes during the experimental years.

94 Cooke, Lez. 2015, *British Television Drama: A History*, London: BFI, 8.

95 Thumim, *Inventing Television Culture*, 47–9.

96 BBC L2/5/2 Memo from Mary Adams to Norman Collins regarding staffing 18 June 1948; BBC L2/5/2 Memo from Mary Adams to Norman Collins regarding understaffing in her department 13 April 1949.

97 BBC L2/5/2 Letter from George Barnes to Norman Collins about Talks Department 1 December 1949.

98 BBC L2/5/2 Norman Collins' report on Mary Adams and Talks Department 16 November 1949.

99 'New Post for West Leeds M.P.'s Wife' (1937), *Leeds Mercury*, January: 6, 6.

100 'Television Appointment' (1937), *Nottingham Journal*, January: 11 5.

101 'The New Order–"Voices Off"' (1937), *Radio Times Television Supplement*, September: 10, 21.

102 'Talks about Many Things' (1939), *Radio Times*, August: 18, 13.

103 'Television News' (1939), *Radio Times*, April: 7, 15.

104 *Radio Times*, 'Television News,' 15.

105 *Rough Island Story* (1939), [Television Programme] BBC.

106 *Radio Times*, 'Television News,' 18.

107 Turnock, Rob. 2007, *Television and Consumer Culture: Britain and the Transformation of Modernity*, London: I.B. Tauris, 16; italics in original.

108 Ibid.

109 Thumin, *Inventing Television Culture*.

Chapter 4

1 Burns, *Television: An International History of the Formative Years*, 573.

2 NBC and CBS were widely reported to have signed some women in television roles during the late 1930s. Women such as Dorothy Knapp, Frances Buss, Thelma A. Prescott and Betty Goodwin were contracted for television at certain points. Others such as Helen Sioussat, Martha Rountree and Gertrude Berg moved between radio and television. Others worked with particular television stations that had experimental licences such as WBKB in Chicago and DuMont's WABD in New York (from 1940). In order to maintain production, some stations recruited women to work mainly in radio, but also on the experimental television productions.

3 Staff magazine *NBC Transmitter* was published in the 1930s and 1940s and distributed nationally to places where NBC or affiliated stations operated. This was followed by *NBC Chimes* in the 1940s and 1950s.

4 Milkman, Ruth. 2013, 'Redefining "Women's Work": The Sexual Division of Labor in the Auto Industry during World War II.' In Ed. Nacy F. Cott. *Industrial Wage Work*, Muchich: K.G. Saur, 508.

5 Christopher Sterling and John Michael Kittross's *Stay Tuned: A History of American Broadcasting* (2001), Gary Edgerton's *The Columbia History of American Television* (2007), David Weinstein's *The Forgotten Network: DuMont and the Birth of American Broadcasting* (2006), James Baughman's *Same Time, Same Station: Creating American Television, 1948-1961* (2007), William Boddy's *Fifties Television: The Industry and*

Its Critics (1993) and Erik Barnouw's trilogy on American broadcasting history (1966, 1968, 1970) pay little attention to the work of women.

6   Women's Bureau. 1947, *Women in Radio*, Washington, DC: United States Department of Labor.

7   Carter, Sue. 1998, '"Women Don't Do News": Fran Harris and Detroit's Radio Station WWJ.' *Michigan Historical Review*, 24 (2): 77–87.

8   St John, Jacqueline D. 1978, 'Sex Role Stereotyping in Early Broadcast History: The Career of Mary Margaret McBride.' *Frontiers: A Journal of Women Studies*, 3 (3): 31–8.

9   Lavin, Marilyn. 1995, 'Creating Consumers in the 1930s: Irna Phillips and the Radio Soap Opera.' *Journal of Consumer Research*, 22 (1): 75–89.

10   Hilmes, *Radio Voices*; Hilmes, Michele. 2013, *Only Connect: A Cultural History of Broadcasting in the United States*, Wadsworth: Cengage Learning.

11   Halper, *Invisible Stars*.

12   Ibid., 116.

13   O'Dell, *Women Pioneers in Television*.

14   Ibid., 5.

15   *I Love Lucy* (1951–7), [Television Programme] CBS.

16   *As the World Turns* (1956–2010), [Television Programme] CBS.

17   O'Dell, Cary. 2013, *June Cleaver Was a Feminist!: Reconsidering the Female Characters of Early Television*, Jefferson, MC: McFarland & Company.

18   Hill, Erin. 2016, *Never Done: A History of Women's Work in Media Production*, New Brunswick, NJ: Rutgers University Press.

19   Levine, Elana. 2020, *Her Stories: Daytime Soap Opera and US Television History*, Durham, NC: Duke University Press.

20   Conway, *The Origins of Television News in American*, 218.

21   Conway, Mike. 2007, 'A Guest in Our Living Room: The Newscaster Before the Rise of the Dominant Anchor.' *Journal of Broadcasting & Electronic Media*, 51 (3): 457–78.

22   Hosley, David H. & Yamada, Gayle K. 1987, *Hard News: Women in Broadcast Journalism*, Westport: Connecticut: Greenwood Press.

23   Hosley and Yamada suggest that Frederick send samples of her interview tapes to Murrow who, in turn, wrote a note that said that her work was fine but that CBS didn't have a requirement for women at that time. While Hosley and Yamada note that Murrow had a strong record of employing women, they state the CBS was not seeking to employ more women.

24 Spaulding, Stacy. 2005, 'Lisa Sergio's "Column of the Air": An
   Examination of the Gendered History of Radio (1940–1945).' *American
   Journalism*, 22 (1): 35–60.

25 Ozmun, David. 2008, 'Opportunity Deferred: A 1953 Case Study of a
   Woman Working in Network Television News.' *Journal of Broadcasting &
   Electronic Media*, 52, 1–15.

26 Corrigan, Thomas. F. 2018, 'Making Implicit Methods Explicit: Trade
   Press Analysis in the Political Economy of Communication.' *International
   Journal of Communication*, 12, 2760.

27 Caldwell, John T. 2008, *Production Culture: Industrial Reflexivity and
   Critical Practice in Film and Television*, Durham, NC: Duke University Press.

28 *NBC Transmitter* published a number of stories on female employees
   who stepped into male roles or undertook jobs typically done by men. A
   September 1942 issue noted that station WTAR now employed a woman
   as control operator as 'loss of male personnel to the Armed Forces had
   caused several stations to seek trained women for the technical jobs'. A
   February 1943 issue, for example, profiled the 'First Girl Sound Effects
   Man,' Betty Boyle, who graduated from NBC's sound effect class in
   Hollywood. See *NBC Transmitter*. 1942. 'Girl at the Switch.' 8 (5): 5; *NBC
   Transmitter*. 1943. 'First Girl Sound Effects 'Man.' 8 (9): 1.

29 US House. Federal Radio Commissions. Seventieth Congress, second
   session. 1929. 'A bill continuing the powers and authority of the Federal
   Radio Commission under The Radio Act of 1927, and for Other
   Purposes.' Washington: Government Printing Office, 722.

30 Slotten, Hugh R. 2000, *Radio and Television Regulation: Broadcast
   Technology in the United States, 1920–1960*, London: Johns Hopkins
   University Press, 72.

31 Ibid.

32 Ibid., 79.

33 von Schilling, James A. 2013, *The Magic Window: American Television
   1939–1953*, New York: Haworth Press.

34 Federal Communications Commission. 1980, *New Television Networks:
   Entry, Jurisdiction, Ownership and Regulation, Volume 1*, Network Inquiry
   Special Staff, 26.

35 United States Congress. Senate Committee on Commerce. 1974,
   *Broadcast License Renewal Act: Hearings Before the Subcommittee on
   Communications of the Committee on Commerce, United States Senate,*

Ninety-third Congress, Second Session, Washington: Government Printing Office, 905.

36 Slotten, *Radio and Television Regulation*, 81.

37 Burns, *Television: An International History of the Formative Years*, 573.

38 Brinson, Susan L. 2007, 'Developing a Television Genre: Table Talk with Helen Sioussat.' *Journal of Broadcasting & Electronic Media*, 51 (3): 418.

39 FCC, *New Television Networks: Entry, Jurisdiction, Ownership and Regulation, Volume 1*, 68.

40 Ibid., 75–6.

41 Slotten, *Radio and Television Regulation*, 81.

42 FCC, *New Television Networks: Entry, Jurisdiction, Ownership and Regulation, Volume 1*, 77.

43 'What Not Column! How Much Do You KNOW?' (1923), *The Northwestern Bulletin*, October: 27, 4.

44 Paxman, Andrew. 2018, 'The Rise of U.S. Spanish-Language Radio.' *Journalism History*, 44 (3): 174–86.

45 Vaillant, Derek W. 2002, 'Sounds of Whiteness: Local Radio, Racial Formation, and Public Culture in Chicago, 1921–1935.' *American Quarterly*, 54 (1): 25–6.

46 *The All-Negro Hour* (1929–35), [Radio Programme] WSBC.

47 Grame, Theodore C. 1980, *Ethnic Broadcasting in the United States*, American Folklore Center: Library of Congress.

48 *The Ethel Waters Show* (1939), [Radio Programme] WX2BS.

49 'Today on the Radio' (1939), *New York Times*, June: 14, 32; 'Today on the Radio' (1939), *New York Times*, June: 16, 46.

50 'Orchids and Peanuts: A Radio Star's Story' (1935), *Detroit Free Press*, June: 30, 11; Cuthbert, Margaret. 1938, 'How to Get into Radio.' *The Key*, 55 (2): 179.

51 Edgerton, Gary. 2010, *The Columbia History of American Television*, New York: Columbia University Press, 53.

52 *Listen.* 1937, 'NBC Newsnames.' September 1937, 1 (2): 7.

53 Cuthbert, 'How to Get into Radio,' 179.

54 'NBC Proves Television Practical' (1937), *Modern Mechanix*, March, 48.

55 'One Minute Interview' (1937), *Radio Daily*, October: 14, 6.

56 'NBC Televises Fashions' (1937), *Broadcasting*, December: 1, 17.

57 Ibid., 17.

58 'Women's Place in Radio Advertising' (1939), *Broadcasting*, July: 1, 48; 52; 62–3.

59 Ibid.

60 'Women in Television' (Spring 1945), *Televisor*, 23.

61 Women who worked in production or technical roles were suggested to be suited to work on women's interest programmes, thus, suggesting that they would not be able to or not competent at work on general interest programmes. This same claim was not made of male production and technical workers.

62 'NBC Televises Fashions' (1937), *Broadcasting*, December: 1, 17.

63 'NBC Looks to the Women's Interest' (1939), *Radio News*, March, 14.

64 Alley, Robert S. & Brown, Irby B. 2001, *Women Television Producers: Transformation of the Male Medium*, Rochester, NY: University of Rochester Press, 5.

65 Swift, Lela. 1987, *Studio One Video History: Lela Swift Interviewed by Loring Mandel*, Museum of Television and Radio. [video]. Lela Swift went on to direct some of the most iconic daytime television programmes including *Dark Shadows* (ABC, 1966–71) and *Ryan's Hope* (ABC, 1975–89).

66 Murray, Matthew. 2002, 'Matinee Theater: Difference, Compromise and the 1950s Daytime Audience.' In Ed. Janet Thumim. *Small Screen, Big Ideas: Television in the 1950s*, London: I.B. Tauris, 135.

67 Boddy, *Fifties Television*, 20.

68 *NBC Transmitter*. 1939, 'Names in the News.' 5 (1): 4.

69 'Television's First Woman Program Director' (1939), *Broadcasting*, February: 1, 84.

70 'Television's First Roadshow Proves a Hit' (1939), *Broadcasting*, February: 15, 2.

71 *NBC Transmitter*. 1939, 'Names in the News.' 5 (1): 4.

72 'Television Holds the Spotlight' (1939), *Short Wave Television*, May, 6.

73 'The Radio Month' (1939), *Radio Craft*, May, 646.

74 *Girl about Town* (29 March 1939), [Television Broadcast] Producer Thelma Prescott, New York: NBC.

75 *NBC Transmitter*. 1939, 'Miscellaneous.' 5 (10): 5.

76 Brochure. Circa. 1940, 'Thelma A. Prescott Discusses Television.' Jacques & Co Inc.

77 O'Dell, Cary. 2017, 'Thelma Prescott, Television's First Female Director/ Producer.' Available at: https://blogs.loc.gov/now-see-hear/2017/05/ thelma-prescott-televisions-first-female-producerdirector/ [Accessed on 12 December 2017].

78 'Television Director Foresees Women Taking Role in Industry' (1940), *Ogdensburg Journal*, April: 20, 3.

79 Ibid., 3.

80 'Personal Notes' (1939), *Broadcasting*, March: 1, 35.

81 *Girl of the Week/Sportswoman of the Week* (1948), [Television Programme]. New York: NBC.

82 'Radio and Television Program Reviews: Girl of the Week' (1948), *Billboard*, September: 25, 8.

83 Irvin, Richard. 2018, *The Early Shows: A Reference Guide to Network and Syndicated Prime-Time Television Series from 1944 to 1949*, Albany, Georgia: BearManor Media.

84 'Television Reviews: Girl about Town' (1948), *Variety*, September: 15, 30.

85 'Sterling Series Packages Pilots and Recoups Losses' (1952), *Billboard*, June: 14, 35.

86 Ibid., 35.

87 Sioussat, Helen. 1943, *Mikes Don't Bite*, New York: L.B. Fischer.

88 *Table Talk with Helen Sioussat* (1941–2), [Television Programme] CBS.

89 Brinson, 'Developing a Television Genre,' 419.

90 Smith, Roger P. 2002, *The Other Face of Public Television: Censoring the American Dream*, New York: Algora Publishing.

91 Hosley & Yamada, *Hard News*.

92 Ibid.

93 Ibid., 45.

94 Ibid., 45–6.

95 Brinson, 'Developing a Television Genre,' 413.

96 Ibid.

97 WCBW daily announcement, quoted in Conway, *The Origins of Television News in America*, 78.

98 Brinson, 'Developing a Television Genre,' 412.

99 Ibid., 416.

100 'Demo's Whopping 23.5 Payoff On Stevenson Saturation TV "Sell"' (1956), *Variety*, September: 19, 43.

101 Conway, *The Origins of Television News in American*, 217.

102 Conway, Mike. 2007, 'Before the Bloggers: The Upstart News Technology of Television at the 1948 Political Conventions.' *American Journalism*, 24 (1): 33–58.

103 Sioussat quoted in Hosley & Yamada, *Hard News*, 43.

104 Ibid.

105 Sioussat, *Mikes Don't Bite*.

106 Ibid.

107 Ibid., 250.

108 Ibid., 244.

109 Sioussat quoted in Hosley & Yamada, *Hard News*, 46.

110 Buss Buch, Frances. 2005, *Archive of American Television: Frances Buss Buch Interview*. Available at: http://www.emmytvlegends.org/interviews/people/frances-buss-buch# [Accessed on 12 December 2017].

111 Ibid.

112 'Meet Frances Buss- Another Do-It-Yourself American!' (1950), *The Post Standard*, February: 6, 5.

113 *Missus Goes a Shopping* (1941–9), [Television Programme] CBS.

114 'Reviews: CBS' (1945), *Billboard*, April: 14, 10.

115 'Stem Shot-Calles Tele-Rated' (1946), *Billboard*, June: 15, 19.

116 *They Were There* (1944), [Television Programme] CBS.

117 'Television Review: They Were There' (1944), *Variety*, October: 25, 40.

118 Buss, *Archive of American Television*.

119 Swift quoted in Englert, Angela. 2016, *The Long, Dark Shadow of Lela Swift*. Available at: http://theculturalgutter.com/horror/the-long-dark-shadow-of-lela-swift.html [Accessed on 12 December 2017].

120 Buss, *Archive of American Television*.

121 'Television Reviews: Vanity Fair' (1949), *Variety*, September: 24, 13.

122 *NBC Transmitter*. 1943, 'First Girl Sound Effects "Man".' 8 (9): 2; *NBC Transmitter*. 1943, 'Summer Institute Achieves Jobs for All '42 Graduates.' 8 (11): 6.

123 *NBC Transmitter*. 1944, 'Survey Shows Vast Number of Replacements by Women.' 9 (7): 14.

124 'Women Take Charge for War Duration of Balaban & Katz Station' (1942), *Broadcasting*, October: 26, 66.

125 O'Dell, Cary. 2000, 'A Station of Their Own: The Story of the Women's Auxiliary Television Technical Staff (WATTS) in World War II Chicago.' *Television Quarterly*, 30 (2): 58.

126 Ibid., 62.

127 The women were recruited throughout the period between 1942 and 1945, with some such as Beulah Zachary joining WBKB in January 1945. *Variety*. 1945. 'In Chicago … ' 17 January 1945, 30.

128 Ibid., 64.

129 'Television Review: Balaban & Katz Television' (1945), *Variety*, February: 7, 38.

130 *Bright Star Shining* (1945), [Television Programme] WBKB.

131 *Variety*, 'Television Review: Balaban & Katz Television,' 37.

132 'Television Reviews: B&K Television' (1944), *Billboard*, April: 15; 7, 32.

133 'Reviews: Balaban & Katz' (1945), *Billboard*, July: 7, 14.

134 *X Marks the Spot* (1944–6), [Television Programme] WBKB.

135 'Review: Balaban & Katz' (1945), *Billboard*, July: 28, 13.

136 Zachary, Beulah (Producer). (18 September 1946), *Music for You* [Television Broadcast]. Chicago: WBKB.

137 'Music for You' (1946), *Billboard*, September: 28, 13.

138 'Radio $$ Lure B&K, Applies for FM Outlet' (1945), *Variety*, November: 17, 8.

139 '"High School" Treatment of Chi Video Production Sizzles Pros' (1946), *Variety*, May: 15, 35.

140 Ibid.

141 Ibid.

142 'Chi Video, Slowed by the War, Begins to Roll; Nets, Indies Push Plans' (1946), *Variety*, January: 9, 119; 122.

143 Gomery, Douglas. 2008, *A History of Broadcasting in the United States*, Oxford: Blackwell, 118.

144 WBKB Weekly programme report, 28 January–1 February produced by Reinald Werrenrath, 8 February 1946, 1. Available at: http://www.earlytelevision.org/pdf/programming_wbkb.pdf [Accessed on 16 September 2018].

145 Ibid., 2.

146 WBKB memo from Reinald Werrenrath outlining studio organization, 9 February 1946. Available at: http://www.earlytelevision.org/pdf/programming_wbkb.pdf [Accessed on 16 September 2018], p. 1.

147 'WBKB Video Changes' (1945), *Broadcasting*, September: 9, 50; 'WBKB's Staffers Join B&K Links to Aid in Video' (1945), *Billboard*, September: 1, 12.

148 'Dave Crandell, Video Vet, Joins WBKB as Producer' (1946), *Billboard*, January: 26, 13.

149 'WBKB Hires Technicians in Station's Expansion' (1946), *Billboard*, February: 16, 15.

150 O'Dell, 'A Station of Their Own.'

151 'Television Reviews' (1946), *Variety*, April: 3, 30; 'Review: Ladies in Retirement' (1948), *Billboard*, August: 14, 11.

152 In the post-war years, women were occasionally recruited in WBKB. However, there was no further campaign to recruit and train large groups of women. Among the women who did work at the station were: Virginia 'Dinny' Butts who worked on camera and writing; Loraine L Groh, who worked across a variety of technical and supervisory roles between 1943 and 1948; Luckey North who was an announcer in the early 1950s and Lee Phillip who took over from North and following her employment at WBKB went on to develop icon soap operas *The Young and the Restless* and *The Bold and the Beautiful*.

153 Paramount Pictures. 1949, *Looking Ahead with Television*, Peori, IL: National Radio Personalities, 25.

154 Mullen, Megan. 2003, *The Rise of Cable Programming in the United States: Revolution or Evolution?*, Austin, TX: University of Texas Press, 29.

155 Alley, *Horrible Prettiness*, 11.

Chapter 5

1 Brouder, Michael & Brookey, Robert Alan. 2015, 'Twitter and Television: Broadcast Ratings in the Web 2.0 Era.' In Ed. John V. Pavlik. *Digital Technology and the Future of Broadcasting: Global Perspectives*, New York: Routledge; Swiencicki, Mark A. 1999, 'Consuming Brotherhood: Men's Culture, Style and Recreation as Consumer Culture, 1880–1930.' In Ed. Lawrence B. Glickman. *Consumer Society in American History: A Reader*, Ithaca, New York: Cornell University Press, 207–40.

2 Igo, Sarah. 2008, *The Averaged American: Surveys, Citizens, and the Making of a Mass Public*, Cambridge, MA: Harvard University Press.

3 Veblen, Thorstein. 1899/2016, *The Theory of the Leisure Class*, Dover Thrift Edition. Mineola, NY: Dover Publications; Reekie, Gail. 1993,

Temptations: Sex, Selling and the Department Store, Sydney: Allen & Unwin.

4   Felski, *The Gender of Modernity*.

5   Schwarzkopf, Stefan. 2016, 'In Search of the Consumer: The History of Market Research from 1890 to 1960.' In Eds. D.G. Brian Jones & Mark Tadajewski. *The Routledge Companion to Marketing History*, London: Routledge, 61–84.

6   Ibid.

7   de Grazia, Victoria. 1996, 'Introduction.' In Eds. Victoria de Grazia & Ellen Furlough. *The Sex of Things: Gender and Consumption in Historical Perspective*, Berkeley, CA: University of California Press, 3.

8   Ibid., 7.

9   Bristor, Julia M. & Fischer, Eileen. 1993, 'Feminist Thought: Implications for Consumer Research.' *Journal of Consumer Research*. 19, March, 523.

10 Ibid.

11 Ibid., 524.

12 Ibid., 528–9.

13 Lears, Jackson. 1994, *Fables of Abundance: A Cultural History of Advertising in America*, New York: Basic Books.

14 McGovern, Charles F. 2009, *Sold American: Consumption and Citizenship, 1890–1945*, Chapel Hill: The University of North Carolina Press, 32.

15 Schwarzkopf, Stefan. 2011, 'The Statisticalization of the Consumer in British Market Research, *c.* 1920–1960: Profiling a Good Society.' In Eds. Tom Crock & Glen O'Hara. *Statistics and the Public Sphere: Numbers and the People in Modern Britain, c. 1800–2000*, London: Routledge, 146.

16 McGovern, *Sold American: Consumption and Citizenship, 1890–1945*, 33.

17 Peiss, Kathy. 1998, 'American Women and the Making of Modern Consumer Culture.' *The Journal for Multi Media History*, 1 (1). Available at: https://www.albany.edu/jmmh/vol1no1/peiss-text.html [Accessed on 12 December 2017].

18 McGovern, *Sold American: Consumption and Citizenship, 1890–1945*, 37.

19 Marchand, Roland. 1985, *Advertising the American Dream: Making Way for Modernity, 1920–1940*, Berkeley, CA: University of California Press, 63–5.

20 Ibid.

21 Rosa-Salas, Marcel. 2019, 'Making the Mass White: How Racial
    Segregation Shaped Consumer Segmentation.' In Eds. Guillaume D.
    Johnson, Kevin D. Thomas & Anthony Kwame Harrison. *Race in the
    Marketplace: Crossing Critical Boundaries*, Basingstoke: Palgrave.

22 Ibid., 25.

23 Marchand, *Advertising the American Dream: Making Way for Modernity,
    1920–1940*, 66.

24 Scanlon, Jennifer. 1995, *Inarticulate Longings: The Ladies' Home Journal,
    Gender, and the Promises of Consumer Culture*, London and New York:
    Routledge, 6–7.

25 Weil Davis, Simone. 2000, *Living Up to the Ads: Gender Fictions in the
    1920s*, Durham, NC: Duke University Press, 186.

26 McGovern, *Sold American: Consumption and Citizenship, 1890–1945*, 37.

27 Ibid.

28 Lavin, 'Creating Consumers in the 1930s: Irna Phillips and the Radio
    Soap Opera,' 78.

29 McGovern, *Sold American: Consumption and Citizenship, 1890–1945*, 48.

30 Ibid.

31 Gale, Harlow. 1900, 'On the Psychology of Advertising.' *Psychological
    Studies*, 1: 39–69.

32 Starch, Daniel. 1914, *Advertising: Its Principles, Practice, and Technique*,
    Chicago, IL: Foresman and Co.

33 Stewart, David W. 2010, 'The Evolution of Market Research.' In Eds.
    Pauline Maclaran, Michael Saren, Barbara Stern & Mark Tadajewski. *The
    SAGE Handbook of Marketing Theory*, London: SAGE, 75–8.

34 Ibid., 76.

35 Ward, Douglas. 2009, *A New Brand of Business: Charles Coolidge
    Parlin, Curtis Publishing Company, and the Origins of Market Research*,
    Philadelphia, PA: Temple University Press.

36 Stivers, Richard. 2001, *Technology as Magic: The Triumph of the Irrational*,
    New York: Continuum, 88.

37 Ibid., 232.

38 Parlin, Charles Coolidge. 1914, *The Merchandising of Textiles*,
    Philadelphia, PA: Curtis Publishing Co.

39 Johnston, Patricia A. 2000, *Real Fantasies: Edward Steichen's Advertising
    Photography*, Berkeley and Los Angeles, CA: University of California
    Press, 162–3.

40 Ibid.

41 Williams Rutherford, Janice. 2010, *Selling Mrs. Consumer: Christine Frederick & the Riske of Household Efficiency*, Athens and London: University of Georgia Press.

42 Darnovsky, Marcy. 1996, *The Green Challenge to Consumer Culture: The Movement, the Marketers, and the Environmental Imagination*, PhD Dissertation. Santa Cruz: University of California, 165.

43 Hollingworth, Harry. 1913, *Advertising and Selling: Principles of Appeal and Response*, New York and London: D. Appleton and Co.

44 Williams Rutherford, *Selling Mrs. Consumer*, 147.

45 Frederick, *Selling Mrs. Consumer*, 17.

46 Ibid., 19.

47 Ibid., 90.

48 Ibid., 19.

49 Ibid., 63.

50 Ibid., 23.

51 Ibid., 24.

52 Ibid.

53 Reilly, William J. 1929, *Marketing Investigations*, New York: The Ronald Press Company.

54 Igo, *The Averaged American*, 5.

55 Hurwitz, Donald. 1983, *Audience Research in American Broadcasting: The Early Years*, August, Oregon, US: Association for Education in Journalism and Mass Communication Annual Convention.

56 Beville, Hugh Malcolm. 1988, *Audience Ratings: Radio, Television, and Cable*, Hove and London: Lawrence Erlbaum Associates, 3.

57 'Bakelite- Faultless Servant of Radio's Invisible Audience' (March 1925), *Radio Age*, 53.

58 'How Big Is the Radio Audience?' (September 1923), *Wireless Age*, 23.

59 Bogart, Leo. 1966, 'Is It Time to Discard the Audience Concept?' *Journal of Marketing*, 30 (1): 48.

60 Craig, Steve. 2010, 'Daniel Starch's 1928 Survey: A First Glimpse of the U.S. Radio Audience.' *Journal of Radio & Audio Media*, 17 (2): 182–94.

61 Meehan, Eileen. 1990, 'Why We Don't Count: The Commodity Audience.' In Ed. Patricia Mellencamp. *Logics of Television: Essays in*

Cultural Criticism, Bloomington and Indianapolis: Indiana University Press, 124.

62 Meehan, Eileen. 1993, 'Heads of Households and Ladies of the House: Gender, Genre, and Broadcast Ratings, 1929–1990.' In Eds. William Samuel Solomon & Robert Waterman McChesney. *Ruthless Criticism: New Perspectives in U.S. Communication History*, Minneapolis and London: University of Minnesota Press.

63 Radio as an Advertising Medium. 1927, *American Radio History*. Available at: http://www.americanradiohistory.com/Archive-Ratings/Crossley%20Nov1927.pdf [Accessed on 12 December 2017].

64 Meehan, 'Heads of Households and Ladies of the House.'

65 Kellogg, H.D. & Walters, A.G. 'How to Reach Housewives Most Effectively' (1932), *Broadcasting*, April: 15; 7 & 30.

66 Ibid.

67 Ibid.

68 Beville, *Audience Ratings*, 29.

69 Ibid.

70 Lumley, Frederick. 1934, *Measurement in Radio*, Columbus, OH: Ohio State University, 190.

71 Ibid., 204.

72 Dygert, Warren. 1939, *Radio as an Advertising Medium*, New York: McGraw Hill, 81.

73 Hettinger, Herman S. 1933, *A Decade of Radio Advertising*, Chicago, IL: University of Chicago Press.

74 Ibid., 208.

75 Ibid.

76 Ibid., 226.

77 Cantril, Hadley & Allport, Gordon M. 1935, *The Psychology of Radio*, New York and London: Harper & Bros.

78 Ibid., 93.

79 Ibid., 94.

80 Ibid.

81 Spaulding, Stacy. 2005, 'Did Women Listen to News? A Critical Examination of Landmark Radio Audience Research (1935–1948).' *Journalism & Mass Communication Quarterly*, 82 (1): 44–61.

82 Ibid., 51.

83 Cantril, Hadley & Allport, Gordon, *The Psychology of Radio*, 92.

84 Haney, David Paul. 2008, *The Americanization of Social Science: Intellectuals and Public Responsibility in the Postwar United States*, Philadelphia, PA: Temple University Press, 60.

85 Glander, Timothy. 1999, *Origins of Mass Communications Research during the American Cold War: Educational Effects and Contemporary Implications*, London: Lawrence Erlbaum Associates, 123.

86 Lazarsfeld, Paul F. 1940, *Radio and the Printed Page*, New York: Duell, Sloan and Pearce.

87 Ibid., 40.

88 Ibid., 164.

89 Stanton, Frank N. & Lazarsfeld, Paul F. 1944, *Radio Research, 1942–1943*, New York: Duell, Sloan and Pearce.

90 Kaufman, Helen J. 1944, 'The Appeal of Specific Daytime Serials.' In Eds. Frank N. Stanton & Paul F. Lazarsfeld. *Radio Research, 1942–1943*, New York: Duell, Sloan and Pearce.

91 Arnheim, Rudolph. 1944, 'The World of the Daytime Serial.' In Eds. Frank N. Stanton & Paul F. Lazarsfeld. *Radio Research, 1942–1943*, New York: Duell, Sloan and Pearce.

92 Herzog, Herta. 1944, 'What Do We Really Know about Daytime Serial Listeners?' In Eds. Frank N. Stanton & Paul F. Lazarsfeld. *Radio Research, 1942–1943*, New York: Duell, Sloan and Pearce, 11–12.

93 Lazarsfeld, Paul F. & Kendell, Patricia L. 1948, *Radio Listening in America: The People Look at Radio- Again*, New York: Prentice-Hall, 221.

94 Ibid., 14.

Chapter 6

1 Spigel, *Make Room for TV: Television and the Family Ideal in Postwar America*.

2 Gould, Jack. 1948, 'Family Life, 1948 A.T. (After Television).' *The New York Times*, August: 1, 12–13.

3 Arthurs, Jane. 2004, *Television and Sexuality: Regulation and the Politics of Taste*, Maidenhead, Berkshire: Open University Press, 34.

4   Ponce de Leon, Charles L. 2015, *That's the Way It Is: A History of Television News in America*, Chicago, IL: University of Chicago Press, 2015, 5.

5   Gershon, Richard A. 2013, *Media, Telecommunications, and Business Strategy*, London: Routledge, 61.

6   US Senate. Committee on Interstate and Foreign Commerce. 1956, *Television Inquiry*, NBC Exhibit 3. Washington: Government Printing Office, 3055–7.

7   Ponce de Leon, *That's the Way It Is*, 6.

8   Hoerschelmann, Olaf. 2006, *Rules of the Game: Quiz Shows and American Culture*, New York: State University of New York Press, 71.

9   Meyers, Cynthia B. 2011, 'The Problem with Sponsorship in US Broadcasting, 1930s–1950s: Perspectives from the Advertising Industry.' *Historical Journal of Film, Radio and Television*, 31 (3): 355–72.

10   Ibid.

11   Ibid.

12   Baughman, James. 2007, *Same Time, Same Station: Creating American Television, 1948–1961*, Baltimore, MD: Johns Hopkins University Press, 4.

13   Ibid., 7.

14   Meyers, 'The Problem with Sponsorship in US Broadcasting, 1930s–1950s,' 366; Stole, Inger L. 2003, 'Televised Consumption: Women, Advertisers and the Early Daytime Television Industry.' *Consumption, Markets and Culture*, 6 (1): 65–80.

15   Meyers, 'The Problem with Sponsorship in US Broadcasting, 1930s–1950s,' 367.

16   Kingson, Walter K. 1953, 'Measuring the Broadcast Audience.' *The Quarterly of Film, Radio and Television*, 7 (3): 295; Beville, Hugh Malcolm. 1988, *Audience Ratings: Radio, Television, and Cable*, Hove and London: Lawrence Erlbaum Associates, 62.

17   Hand Ludlow, John. 1957, *Television Program Preferences of Listeners in Utah County, Utah*, PhD. Ohio: The Ohio State University.

18   Kingson, 'Measuring the Broadcast Audience,' 296–7.

19   JWT Media Research Department. 1958, 'Manual of TV Basics.' J. Walter Thompson Company Archives. Chicago Office. Information Center Records, 1901–2005 and undated. Box R14. Rare Book, Manuscript, and Special Collections Library, Duke University, Durham, North Carolina.

20   The Schwerin Research Corporation, for example, produced monthly bulletins on research it had undertaken and summarized its findings. In

addition, it reported on other research of audiences as well as on research methodologies. These bulletins were addressed to advertisers and advertising agency executives and were concerned with how broadcasters and advertisers could best reach the audience.

21   Hand, *Television Program Preferences of Listeners in Utah County, Utah*, 26–7.

22   Ibid., 29.

23   Gibson, Robert L. 1946, 'Some Preferences of Television Audiences.' *Journal of Marketing*, 10 (3) January 1946: 289–90.

24   Coffin, Thomas E. 1950, *The Hofstra Study: A Measure of the Sales Effectiveness of Television Advertising*, New York: NBC.

25   Baughman, *Same Time, Same Station*, 196.

26   Ripley, Joseph M. 1955, *Levels of Attention of Women Listeners to Daytime and Evening Television Programs in Columbus, Ohio*, Columbus, OH: The Ohio State University.

27   JWT TV-Radio Research Department. 1958, 'Letter from Sarah Frank of TV-Radio Research Department to Irene Dunne, JWT.' 9 April. J. Walter Thompson Company Archives. John F. Devine Papers. Programme Subseries. Box 19. Folder 3. Rare Book, Manuscript, and Special Collections Library, Duke University, Durham, North Carolina.

28   Cassidy, Marsha F. 2009, *What Women Watched: Daytime Television in the 1950s*, Austin, TX: University of Texas Press, 27.

29   Levine, *Her Stories: Daytime Soap Operas and US Television History*, 35.

30   Ibid., 29.

31   'For Daytime Television See Du Mont' (1949), *Sponsor*, August: 15, 13.

32   Spigel, Lynn. 1989, 'The Domestic Economy of Television Viewing in Postwar America.' *Critical Studies in Mass Communication*, 6 (4): 342.

33   Stole, Inger L. 2003, 'Televised Consumption: Women, Advertisers and the Early Daytime Television Industry.' *Consumption, Markets and Culture*, 6 (1): 70.

34   Spigel, *Make Room for TV*, 78.

35   Levine, *Her Stories: Daytime Soap Operas and US Television History*, 40.

36   Ibid., 38.

37   Ibid., 39–40.

38   '63 Stations to Start Programming at 10' (1951), *Billboard*, August: 6, 6.

39   Ibid.

40 Cassidy, *What Women Watched*, 5.

41 *Search for Tomorrow* (1951–86), [Television Programme] CBS/NBC.

42 *As the World Turns* (1956–2010), [Television Programme] CBS.

43 Cunningham & Walsh had a vested interest in demonstrating strong television viewership since the agency created television advertisements for its clients. For example, Cunningham & Walsh's client Chesterfield sponsored the NBC-produced programmes *Dragnet* and CBS-produced programme *Gunsmoke*.

44 Cunningham & Walsh. 1951, *Videotown IV 1951*, Library of American Broadcasting. University of Maryland.

45 Advertest, 1950, 'Study of Daytime Television Number 2.' 2 (16). NBC Files: Box 193, folder 10.

46 W.R. Simmons and Associates Research, Inc. 1954, *Television's Daytime Profile: Buying Habits and Characteristics of the Audience*, NBC Files: Box 183, folder 5. Wisconsin Historical Society Archives.

47 Stanton, Michael Joseph. 1977, *A History of the Research and Planning Department of the National Broadcasting Company, Incorporated (1931 to 1976)*, PhD Thesis. Ann Arbor, MI: University Microfilms International, 110.

48 Ibid.

49 Cunningham & Walsh. 1949, *Videotown - One Year Later 1948/1949*, Library of American Broadcasting. University of Maryland; Cunningham & Walsh. 1950, *Videotown III 1950*, Library of American Broadcasting. University of Maryland.

50 Cunningham & Walsh. 1951, *Videotown IV 1951*, Library of American Broadcasting. University of Maryland.

51 Ibid., 19.

52 Ibid., 21.

53 Ibid., 21.

54 Cunningham & Walsh. 1952, *Videotown 5 1952*, Library of American Broadcasting. University of Maryland, 5.

55 Ibid.

56 Ibid.

57 Ibid., 8.

58 The report continued with a section comparing TV viewing to radio listening and focused mainly on that of wives, with a section for wives

and a section for 'all people'. It indicated that 10 per cent of wives watched TV in the mornings as compared with 2 per cent the previous year. And 18 per cent watched during the afternoon in 1952 as compared to 10 per cent in 1951 and 73 per cent of wives watched in the evening in 1952 as compared to 71 per cent in 1951. The percentage of the total audience for morning was in 1952 5 per cent for mornings, 15 per cent for afternoons and 70 per cent for evenings.

59 Cunningham & Walsh. 1953, *Videotown 6 1948–1953*, Library of American Broadcasting. University of Maryland, 2.

60 Ibid., 2–3.

61 Ibid., 3.

62 Cunningham & Walsh. 1954, *Videotown 7th Edition 1948–1954*, Library of American Broadcasting. University of Maryland, 3.

63 Housewives, it was noted, 'like the rest of the family reduced their total weekday daytime viewing of TV – morning viewing [was] down from 2 hours 21 minutes per week in 1954 to 1 hour 2 minutes in 1955; afternoon viewing [was] down from 2 hours 42 minutes per week in '54 to 2 hours 21 minutes [in 1955].'

64 Cunningham & Walsh. 1955, *Videotown 8th Edition*, Library of American Broadcasting. University of Maryland, 7.

65 Ibid., 18.

66 'Monochrome TV Levels Off as Color Starts Its Push' (1956), *Broadcasting*, October: 15, 44.

67 Ibid., 46.

68 'Column Switch' (1958), *Sponsor*, December: 13, 22.

69 Ibid., 22 & 38.

70 Advertest, 'Study of Daytime Television (1949).'

71 Advertest, 1950. 'Study of Daytime Television Number 2.' 2 (16). NBC Files: Box 193, folder 10.

72 Advertest, 'Study of Daytime Television (1949),' 15.

73 Ibid., 12.

74 Ibid., 15.

75 Ibid., 9.

76 Ibid., 5.

77 Ibid., 6.

78 Ibid., 7.

79 Ibid., 8.

80   Ibid., 23.

81   Ibid., 13.

82   Ibid., 14.

83   *Okay, Mother* (1948–51), [Television Programme] DuMont.

84   *Needle Shop* (1948–9), [Television Programme] DuMont.

85   *Television Shopper* (1948–50), [Television Programme] DuMont.

86   Advertest, 'Study of Daytime Television (1949),' 18.

87   Advertest, 'Study of Daytime Television Number 2 (1950),' 1.

88   Ibid., 2.

89   Ibid., 11.

90   As William Boddy notes, the networks were in the early years of television reluctant to schedule films in the evening time since they were keen on gaining control over production. Therefore, telefilms were, in the early 1950s, scheduled during the day when there were fewer viewers and when advertising costs were lower. See Boddy, William. 1985, 'The Studios Move into Prime Time: Hollywood and the Television Industry in the 1950s.' *Cinema Journal*, 24 (4): 23–37.

91   Advertest, 'Study of Daytime Television Number 2 (1950),' 8–10.

92   Ibid., 11.

93   *Rumpus Room* (1949–52), [Television Programme] DuMont.

94   'What's New in Research?' (1952), *Sponsor*, March: 10, 50.

95   'What's New in Research?' (1952), *Sponsor*, April: 1, 78.

96   *Arthur Godfrey and His Friends* (1949–59), [Television Programme] CBS.

97   'What's New in Research?' (1952), *Sponsor*, October: 6, 80.

98   Meehan, 'Gendering the Commodity Audience,' 317.

99   Further studies were carried out throughout 1954.

100   NBC Research and Planning Department report on Drugs and Toiletries, 26 October 1954. NBC Files: Box 183, file 12. Wisconsin Historical Society Archives, 1.

101   *Today* (1952–), [Television programme] NBC.

102   *Home* (1954–7), [Television programme] NBC.

103   *Ding Dong School* (1952–6), [Television programme] NBC.

104   *The Pinky Lee Show* (1950–6), [Television programme] NBC.

105   Cassidy, *What Women Watched*; Spigel, 'The Domestic Economy of Television Viewing in Postwar America'; Stole, Inger L. 2003, 'Televised Consumption: Women, Advertisers and the Early Daytime Television Industry.' *Consumption, Markets and Culture*, 6 (1): 65–80.

106 Stole, 'Televised Consumption,' 74.

107 NBC Research and Planning Department report on *Zoo Parade*
     audience, 14 April 1954. NBC Files: Box 183, file 12. Wisconsin
     Historical Society Archives.

108 Ibid.

109 Marvin Baiman, NBC Research and Planning Department, letter to
     Mr. William Asip regarding *Today* audience data for Royal Typewriter,
     24 September 1954. NBC Files: Box 183, file 12. Wisconsin Historical
     Society Archives.

110 Ibid.

111 Marvin Baiman, NBC Research and Planning Department, letter to
     Mr. William Hoffman regarding the occupation of the *Today* audience,
     21 December 1955. NBC Files: Box 183, file 12. Wisconsin Historical
     Society Archives.

112 Ibid.

113 *Meet the Press* (1947–), [Television programme] NBC.

114 Robert, D. Daubenspeck, NBC Research and Planning Department,
     letter to Mr. Mario Kircher of J. Walter Thompson regarding audience
     for Meet the Press, 25 January 1956. NBC Files: Box 183, file 13.
     Wisconsin Historical Society Archives.

115 Hilmes, *Radio Voices*, 154.

116 Hatch, Kristen. 2002, 'Selling Soap: Post-War Television Soap Opera and
     the American Housewife.' In Ed. Janet Thumin. *Small Screens, Big Ideas:
     Television in the 1950s*, London: I.B. Tauris, 36.

Chapter 7

1   Silvey, Robert. 1974, *Who's Listening? The Story of BBC Audience Research*,
    London: Allen & Unwin, 15.

2   Street, Sean. 2000, 'BBC Sunday Policy and Audience Response: 1930–
    1945.' *Journal of Radio Studies*, 7 (1): 165.

3   Ibid., 167.

4   Balnaves, O'Regan & Goldsmith, *Rating the Audience*, 92.

5   Ibid., 94.

6   Moores, Shaun. 2007, *Media/Theory: Thinking about Media and
    Communications*, London and New York: Routledge, 21.

7   Ang, *Desperately Seeking the Audience*, 3.

8   See Meehan, Eileen. 1984, 'Ratings and the Institutional Approach: A
    Third Answer to the Commodity Questions.' *Critical Studies in Mass
    Communication*, 1 (2): 216–25; Napoli, Philip M. 2011, *Audience
    Evolution: New Technologies and the Transformation of Media Audiences*,
    New York: Columbia University Press.

9   Gitelman, Lisa. 2013, *'Raw Data' Is an Oxymoron*, Cambridge, MA: MIT
    Press, 3.

10 Napoli, *Audience Evolution: New Technologies and the Transformation of
    Media Audiences*, 7.

11 Silvey, *Who's Listening? The Story of BBC Audience Research*, 34.

12 Schwarzkopf, Stefan. 2014, 'The Politics of Enjoyment: Competing
    Audience Measurement Systems in Britain, 1950–1980.' In Eds. Jerome
    Bourdon & Cecile Meadal. *Television Audiences around the World:
    Deconstructing the Ratings Machine*, New York: Palgrave Macmillan.

13 See Silvey, *Who's Listening? The Story of BBC Audience Research*; Napoli,
    *Audience Evolution: New Technologies and the Transformation of Media
    Audiences*, 10.

14 See Summerfield, Penny. 2000, 'It Did Me Good in Lots of Ways': British
    Women in Transition from War to Peace.' In Eds. Claire Duchen & Irene
    Bandhauer-Schoffmann. *When the War Was Over: Women, War, and
    Peace in Europe, 1940–1956*, London: Leicester University Press, 13;
    Elliot, B. Jane. 1991, 'Demographic Trends in Domestic Life, 1945–87.' In
    Ed. David Clark. *Marriage, Domestic Life and Social Change: Writings for
    Jacqueline Burgoyne 1944–88*, London: Routledge, 86.

15 Stanton, Andrea L. 2012, 'Television.' In Eds. Andrea L. Stanton,
    Edward Ramsamy, Peter J. Seybolt & Carolyn M. Elliott. *Cultural
    Sociology of the Middle East, Asia, and Africa: An Anthology*, London:
    SAGE, 363.

16 Harper, Richard. 2006, *Inside the Smart Home*, New York: Springer, 70.

17 BBC WAC T23/108 Publicity Report on Television by the Postmaster
    General 31 January 1935.

18 BBC WAC T23/108 Publicity Report on the Television Service from the
    Alexandra Palace by Controller (Engineer) 10 August 1936.

19 BBC WAC T1/6/1 'Viewers and the Television Service: A Report of an
    Investigation of Viewers' Opinions in January 1937', 5 February 1937.

20 Burns, *Television: An International History of the Formative Years*, 592.

21 BBC News. 2005, 'Pre-War Britons "Were Happier." *BBC News Online*, 1 September 2005. Available at: http://news.bbc.co.uk/2/hi/uk_news/wales/4203686.stm [Accessed on 14 May 2018].

22 BBC WAC T1/6/1 'Viewers and the Television Service: A Report of an Investigation of Viewers' Opinions in January 1937', 5 February 1937.

23 Ibid.

24 *Quarter-of-an-hour Meals* (1936), [Television Programme] BBC.

25 *Accidents in the Home* (1937), [Television Programme] BBC.

26 *Demonstration by the Women's League of Health and Beauty* (1937), [Television Programme] BBC.

27 BBC WAC T1/6/1 'Viewers and the Television Service: A Report of an Investigation of Viewers' Opinions in January 1937', 5 February 1937.

28 Wood, Helen. 2015, 'Television- the Housewife's Choice? The 1949 Mass Observation Television Directive, Reluctance and Revision.' *Media History*, 21 (3): 345.

29 BBC WAC R9/9/6 Television Conference: Viewers' Questions Monday 26 June 1939.

30 BBC WAC R9/9/6 Memo: Viewers' Party and Questionnaire 28 June 1939.

31 BBC WAC T1/11 Television Enquiry 1939 Interim Report 4 April 1939.

32 BBC WAC T1/11 LR/75 An Enquiry into Viewers' Opinions on Television Programmes conducted in the first quarter of 1939, June 1939.

33 BBC WAC T1/11 LR/75 An Enquiry into Viewers' Opinions on Television Programmes conducted in the first quarter of 1939: introduction June 1939.

34 Ibid.

35 Ibid.

36 BBC WAC R9/9/1 A Report on the Variety (Light Entertainment) Listening Barometer October–December 1937, 29 November 1937.

37 BBC WAC LR/67 Variety Listening Barometer Interim Report 5 April 1938.

38 BBC WAC LR/67 Variety Listening Barometer Interim Report No. 7. Vaudeville in National and Regional Programmes 14 March 1938.

39 BBC WAC LR/67 Winter Listening Habits: A Report on the First Random Sample Scheme 1 September 1938.

40 Silvey, *Who's Listening? The Story of BBC Audience Research*, 114.

41 Ibid., 115.

42 BBC WAC R9/9/6 A Listener Research Report Daytime Repeats 1942.

43 BBC WAC R9/21 Memo from Maurice Gorham to Robert Silvey 29 January 1946.

44 BBC WAC T1/6/2 Audience Research Memos: Gorham asks DG to reconsider his refusal for television audience research 28 June 1946.

45 BBC WAC T1/6/2 Gorham to Senior Controller on the Solicitation of Correspondence from Viewers 21 August 1946.

46 BBC WAC R9/21 Plans for Viewer Research: Memo from Head of Television Service (Gorham) to Editor, *Radio Times*, 6 September 1946.

47 BBC WAC R9/21 Plans for Viewer Research: Draft Letter and Detail form to send to viewers willing to receive questionnaires 9 April 1946.

48 Ibid.

49 BBC WAC T1/6/2 Audience Research Memos: Director of Administration (Bottomley) writes to Silvey 6 July 1948.

50 BBC WAC T1/6/2 Audience Research Memos: A Listener Research Report – Television: Some Points about the Audience 1 July 1948.

51 Ibid.

52 Briggs, Asa. 1995, *Competition: A History of Broadcasting in the United Kingdom (1955–1974)*, London: Oxford University Press, 230.

53 BBC WAC T1/6/2 Mass Observation Report on Television July 1949.

54 Wood, 'Television – the Housewife's Choice?' 345.

55 Ibid., 345.

56 BBC WAC R9/21 BBC Television Panel Log for Week 3 sample log for viewing week 15 January 1950.

57 BBC WAC R9/21 Methods of Viewer Research Employed by the British Broadcasting Corporation by Robert Silvey 13 December 1950.

58 Ibid., 99–100.

59 BBC WAC R9/21 Silvey memo on report on women's and children's programmes 22 March 1950.

60 BBC WAC R9/21 Memo from Silvey to Controller of Television Programmes 27 November 1950.

61 Irwin, 'What Women Want on Television: Doreen Stephens and BBC Television Programmes for Women, 1953–1964,' 99–122.

Conclusion

1   Erdman, *Blue Vaudeville*; Butsch, Richard. 2000, *The Making of American
    Audiences: From Stage to Television, 1750–1990*, Cambridge: Cambridge
    University Press.

2   Kibler, Alison. 2005, *Rank Ladies: Gender and Cultural Hierarchy in
    American Vaudeville*, Chapel Hill: University of North Carolina Press, 5.

3   Ibid.

4   Mulvey, 'Visual Pleasure and Narrative Cinema'; Doane, Mary Ann. 1991,
    Femme Fatales: Film, Feminism and Psychoanalysis, London: Routledge;
    Mayne, Judith. 2002, *Cinema and Spectatorship*, London: Routledge.

5   Hilmes, 'Desired and Feared: Women's Voices in Radio History.'

6   Ibid., 141–4.

7   Ibid.

8   Mulvey, 'Visual Pleasure and Narrative Cinema.'

9   Ehrick, Christine. 2015, *Radio and the Gendered Soundscape: Women
    and Broadcasting in Argentina and Uruguay, 1930–1950*, New York:
    Cambridge University Press, 17.

10 Sassatelli, Roberta. 2007, *Consumer Culture: History, Theory and Politics*,
    London: SAGE, 45.

11 Asquer, Enrica. 2012, 'Domesticity and Beyond: Gender, Family and
    Consumption in Modern Europe.' In Ed. Frank Trentmann. *The Oxford
    Handbook of the History of Consumption*, Oxford: Oxford University
    Press, 575.

12 Marchand, *Advertising the American Dream: Making Way for Modernity,
    1920–1940*, 66.

13 Ibid.

14 Frederick, *Selling Mrs. Consumer*.

15 Butsch, *The Making of American Audiences*, 196.

16 Lauzen, Martha M. 2019, *Boxed In 2018–19: Women on Screen and behind
    the Scenes in Television*, San Diego State University: Center for the Study
    of Women in Television & Film. Available at: https://womenintvfilm.
    sdsu.edu/wp-content/uploads/2019/09/2018-19_Boxed_In_Report.pdf
    [Accessed on 4 January 2019], p. 3.

17 Ofcom. 2017, Diversity of UK television industry revealed.
    Available at: https://www.ofcom.org.uk/about-ofcom/latest/media/

media-releases/2017/diversity-uk-television-industry [Accessed on 4 January 2019].

18 Arnold, Sarah. 2016, 'Netflix and the Myth of Choice/Participation/ Autonomy.' In Eds. Kevin McDonald & Daniel Smith-Rowsey. *The Netflix Effect: Technology and Entertainment in the 21st Century*, New York: Bloomsbury.

Bibliography

Abramson, Albert. *The History of Television: 1881–1941*. Jefferson, NC: McFarland and Company, 1987.

Abramson, Albert. *The History of Television: 1942–2000*. Jefferson, NC: McFarland and Company, 2003.

Advertest. 'Study of Daytime Television.' *The Television Audience of Today*. 1, no. 1. NBC Files: Box 193, folder 2. Wisconsin Historical Society Archives, 1949.

Agnew, Jeremy. *Entertainment in the Old West: Theater, Music, Circuses, Medicine Shows, Prizefighting and Other Popular Amusements*. Jefferson, NC: McFarland & Co, 2011.

Aldridge, Mark. *The Birth of British Television: A History*. London: Palgrave: Macmillan, 2011.

Allen, Robert Clyde. *Horrible Prettiness: Burlesque and American Culture*. Chapel Hill: University of North Carolina Press, 2000.

Alley, Robert S. & Brown, Irby B. *Women Television Producers: Medium*. Rochester, NY: University of Rochester Press, 2001.

Ang, Ien. *Desperately Seeking the Audience*. London: Routledge, 1991.

Ang, Ien. *Watching Dallas: Soap Opera and the Melodramatic Imagination*. London and New York: Routledge, 1996.

Arceneaux, Noah. 'Department Stores and Television.' *Journalism History*. 43, no. 4 (2018): 219–27.

Arceneaux, Ronald J. 'Noah.' *Department Stores and the Origins of American Radio Broadcasting, 1910–1931*, PhD Dissertation. University of Georgia. 2007. Available at: https://getd.libs.uga.edu/pdfs/arceneaux_ronald_j_200705_phd.pdf. [Accessed on 26 July 2020].

Arnheim, Rudolph. 'The World of the Daytime Serial.' In Eds. Frank N. Stanton & Paul F. Lazarsfeld. *Radio Research, 1942–1943*. New York: Duell, Sloan and Pearce, 1944.

Arnold, Sarah. 'Netflix and the Myth of Choice/Participation/Autonomy.' In Eds. Kevin McDonald & Daniel Smith-Rowsey. *The Netflix Effect: Technology and Entertainment in the 21st Century*. New York: Bloomsbury, 2016.

Arthurs, Jane. *Television and Sexuality: Regulation and the Politics of Taste.* Maidenhead, Berkshire: Open University Press, 2004.

Asquer, Enrica. 'Domesticity and Beyond: Gender, Family and Consumption in Modern Europe.' In Ed. Frank Trentmann. *The Oxford Handbook of the History of Consumption.* Oxford: Oxford University Press, 2012.

Babst, Joacob L. & Tribe, Ivan M. *Beryl Halley: The Life and Follies of a Ziegfeld Beauty, 1897–1988.* Jefferson, NC: McFarland, 2019.

Bailey, Michael. 'The Angel in the Ether: Early Radio and the Construction of the Household.' In Ed. Michael Bailey. *Narrating Media History.* London: Routledge, 2009.

Balnaves, Mark, O'Regan, Tom & Goldsmith, Ben. *Rating the Audience: The Business of Media.* London: Bloomsbury Academic, 2011.

Barnouw, Erik. *A Tower in Babel: A History of Broadcasting in the United States to 1933.* London: Oxford University Press, 1966.

Barnouw, Erik. *The Golden Web: A History of Broadcasting in the United States 1933–1953.* London: Oxford University Press, 1968.

Barnouw, Erik. *The Image Empire: A History of Broadcasting in the United States from 1953.* London: Oxford University Press, 1970.

Baughman, James. *Same Time, Same Station: Creating American Television, 1948–1961.* Baltimore, MD: Johns Hopkins University Press, 2007.

Beauvoir, Simone de. *The Second Sex.* Trans. Constance Border and Sheila Malovany-Chevallier. New York: Random House, 1949/2009.

Becker, Ron. '"Hear-and-See Radio" in the World of Tomorrow: RCA and the Presentation of Television at the World's Fair, 1939–1940.' *Historical Journal of Film, Radio and Television.* 21, no. 4 (2001): 361–78.

Bennett, James. *Television Personalities: Stardom and the Small Screen.* London: Routledge, 2010.

Beville, Hugh Malcolm. *Audience Ratings: Radio, Television, and Cable.* Hove and London: Lawrence Erlbaum Associates, 1988.

Boddy, William. 'The Studios Move into Prime Time: Hollywood and the Television Industry in the 1950s.' *Cinema Journal.* 24, no. 4 (1985): 23–37.

Boddy, William. *Fifties Television: The Industry and Its Critics.* Chicago, IL: University of Illinois Press, 1993.

Bogart, Leo. 'Is It Time to Discard the Audience Concept?' *Journal of Marketing.* 30, no. 1 (1966): 47–54.

Briggs, Asa. *The Birth of Broadcasting: The History of Broadcasting in the United Kingdom (1896–1927).* London: Oxford University Press, 1961.

Briggs, Asa. *The Golden Age of Wireless: The History of Broadcasting in the United Kingdom (1926–1939)*. 1965. London: Oxford University Press, 1995.

Briggs, Asa. *The War of Words: The History of Broadcasting in the United Kingdom (1939–1945)*. London: Oxford University Press, 1970.

Briggs, Asa. *Sound and Vision: The History of Broadcasting in the United Kingdom (1945–1955)*. London: Oxford University Press, 1979.

Briggs, Asa. *Competition: A History of Broadcasting in the United Kingdom (1955–1974)*. London: Oxford University Press, 1995.

Brinson, Susan L. 'Developing a Television Genre: Table Talk with Helen Sioussat.' *Journal of Broadcasting & Electronic Media*. 51, no. 3 (2007): 410–23.

Bristor, Julia M. & Fischer, Eileen. 'Feminist Thought: Implications for Consumer Research.' *Journal of Consumer Research*. 19 (March 1993): 518–36.

Brouder, Michael & Brookey, Robert Alan. 'Twitter and Television: Broadcast Ratings in the Web 2.0 Era.' In Ed. John V. Pavlik. *Digital Technology and the Future of Broadcasting: Global Perspectives*. New York: Routledge, 2015.

Burns, R. W. *John Logie Baird: Television Pioneer*. London: The Institution of Electronic Engineers, 2000.

Burns, Russell W. *British Television: The Formative Years*. London: The Institution of Electronic Engineers, 1986.

Burns, Russell W. *Television: An International History of the Formative Years*. London: The Institution of Electronic Engineers, 1998.

Burns, Russell W. *Communications: An International History of the Formative Years*. London: The Institution of Electronic Engineers, 2004.

Buss Buch, Frances. *Archive of American Television: Frances Buss Buch Interview*. 2005. Available at: http://www.emmytvlegends.org/interviews/people/frances-buss-buch# [Accessed on 12 December 2017].

Butsch, Richard. *The Making of American Audiences: From Stage to Television, 1750–1990*. Cambridge: Cambridge University Press, 2000.

Caldwell, John T. *Production Culture: Industrial Reflexivity and Critical Practice in Film and Television*. Durham, NC: Duke University Press, 2008.

Cantril, Hadley & Allport, Gordon M. *The Psychology of Radio*. New York and London: Harper & Bros, 1935.

Carter, Cynthia, Branston, Gill & Allan, Stuart. *News, Gender, and Power*. London: Taylor & Francis, 1998.

Carter, Sue. "'Women Don't Do News": Fran Harris and Detroit's Radio Station WWJ.' *Michigan Historical Review.* 24, no. 2 (1998): 77–87.

Casey, Emma & Martens, Lydia. *Gender and Consumption: Domestic Cultures and the Commercialisation of Everyday Life.* London and New York: Routledge, 2012.

Cassidy, Marsha F. *What Women Watched: Daytime Television in the 1950s.* Austin, TX: University of Texas Press, 2009.

Cockburn, Cynthia. *Machinery of Dominance: Women, Men and Technical Know-How.* London: Pluto Press, 1985.

Cockburn, Cynthia. 'The Circuit of Technology: Gender, Identity and Power.' In Eds. Roger Silverstone & Eric Hirsch. *Consuming Technologies: Media and Information in Domestic Spaces.* London: Routledge, 1992.

Coffin, Thomas E. *The Hofstra Study: A Measure of the Sales Effectiveness of Television Advertising.* New York: NBC, 1950.

Conway, Mike. 'A Guest in Our Living Room: The Newscaster before the Rise of the Dominant Anchor.' *Journal of Broadcasting & Electronic Media.* 51, no. 3 (2007): 457–78.

Conway, Mike. 'Before the Bloggers: The Upstart News Technology of Television at the 1948 Political Conventions.' *American Journalism.* 24, no. 1 (2007): 33–58.

Conway, Mike. *The Origins of Television News in American: The Visualizers of CBS in the 1940s.* New York: Peter Lang, 2009.

Cooke, Lez. *British Television Drama: A History.* London: BFI, 2015.

Cornell, Paul, Day, Martin & Topping, Keith. *The Classic British Telefantasy Guide.* London: Hachette UK, 2004.

Corrigan, Thomas. F. 'Making Implicit Methods Explicit: Trade Press Analysis in the Political Economy of Communication.' *International Journal of Communication.* 12 (2018): 2751–72.

Cott, Nancy. *The Grounding of Modern Feminism,* New Haven, CN: Yale University Press, 1987.

Cowan, Ruth Schwartz. *More Work for Mother: The Ironies of Household Technology from the Open Hearth to the Microwave.* New York: Basic Books, 1983.

Craig, Steve. 'Daniel Starch's 1928 Survey: A First Glimpse of the U.S. Radio Audience.' *Journal of Radio & Audio Media.* 17, no. 2 (2010): 182–94.

Crook, David. 'School Broadcasting in the United Kingdom: An Exploratory History.' *Journal of Educational Administration and History.* 39, no. 3 (2007): 217–26.

Cunningham & Walsh. *Videotown- One Year Later 1948/1949*. Library of American Broadcasting. University of Maryland, 1949.

Cunningham & Walsh. *Videotown III 1950*. Library of American Broadcasting. University of Maryland, 1950.

Cunningham & Walsh. *Videotown IV 1951*. Library of American Broadcasting. University of Maryland, 1951.

Cunningham & Walsh. *Videotown 5 1952*. Library of American Broadcasting. University of Maryland, 1952.

Cunningham & Walsh. *Videotown 6 1948–1953*. Library of American Broadcasting. University of Maryland, 1953.

Cunningham & Walsh. *Videotown 7th Edition 1948–1954*. Library of American Broadcasting. University of Maryland, 1954.

Cunningham & Walsh. *Videotown 8th Edition*. Library of American Broadcasting. University of Maryland, 1955.

Currie, Tony. *British Television 1930–2000*. Second Edition. Devon: Kelly Publications, 2004.

Darnovsky, Marcy. *The Green Challenge to Consumer Culture: The Movement, The Marketers, and the Environmental Imagination*. PhD. University of California, Santa Cruz, 1996.

de Grazia, Victoria. 'Introduction.' In Eds. Victoria de Grazia & Ellen Furlough. *The Sex of Things: Gender and Consumption in Historical Perspective*. Berkeley, CA: University of California Press, 1996.

de Leon, Ponce, Charles, L. *That's the Way It Is: A History of Television News in America*. Chicago, IL: University of Chicago Press, 2015.

Doane, Mary Ann. *Femme Fatales: Film, Feminism and Psychoanalysis*. London: Routledge, 1991.

Doughan, David & Gordon, Peter. *Women, Clubs and Associations in Britain*. London: Routledge, 2007.

Douglas, Ann. *The Feminization of American Culture*. New York: Alfred A. Knopf, 1977.

Douglas, Susan J. *Listening In: Radio and the American Imagination*. Minneapolis, MI: University of Minnesota Press, 2004.

Dunlap, Orrin E. *The Outlook for Television*. New York and London: Harper & Brothers Publishers, 1932.

Dyer, Richard. *Heavenly Bodies: Film Stars and Society*. Second Edition. London: Routledge, 2004.

Dygert, Warren. *Radio as an Advertising Medium*. New York: McGraw Hill, 1939.

Edgerton, Gary. *The Columbia History of American Television*. New York: Columbia University Press, 2010.

Ehrick, Chrisine. *Radio and the Gendered Soundscape: Women and Broadcasting in Argentina and Uruguay, 1930-1950*. New York: Cambridge University Press, 2015.

Elliot, B. Jane. 'Demographic Trends in Domestic Life, 1945-87.' In Ed. David Clark. *Marriage, Domestic Life and Social Change: Writings for Jacqueline Burgoyne 1944-88*. London: Routledge, 1991.

Erdman, Andrew L. *Blue Vaudeville: Sex, Morals and the Mass Marketing of Amusement, 1895-1915*. London: McFarland & Company, 2007.

Everson, George. *The Story of Television: The Life of Philo T. Farnsworth*. New York: W. W. Norton, 1949.

Fairclough, Norman & Wodak, Ruth. 'Critical Discourse Analysis.' In Ed. Teun van Dijk. *Discourse as Social Interaction*. London: SAGE, 1997.

Faulkner, Wendy. 'The Technology Question in Feminism: A View from Feminist Technology Studies.' *Women's Studies International Forum*. 24, no. 1 (2001): 79-95.

Felski, Rita. *The Gender of Modernity*. Cambridge, MA: Harvard University Press, 1995.

Fisher, David E. & Fisher, Marshall. *Tube: The Invention of Television*. Berkeley, CA: Counterpoint, 1996.

Foucault, Michel. *The Order of Things: An Archaeology of Human Sciences*. New York: Vintage Books, 1994.

Frederick, Christine McGaffey. *Selling Mrs. Consumer*. New York: The Business Bourse, 1929.

Friedberg, Anne. *Window Shopping: Cinema and the Postmodern*. Berkeley, CA: University of California Press, 1994.

Gale, Harlow. 'On the Psychology of Advertising.' *Psychological Studies*. 1 (1900): 39-69.

Gardner, Viv. 'The Theatre of the Flappers?: Gender, Spectatorship and the "Womanisation" of Theatre 1914-1918.' In Ed. Andrew Maunder. *British Theatre and the Great War, 1914-1919*. London: Palgrave Macmillan, 2015.

Gershon, Richard A. *Media, Telecommunications, and Business Strategy*. London: Routledge, 2013.

Gibson, Robert L. 'Some Preferences of Television Audiences.' *Journal of Marketing*. 10, no. 3 (January 1946): 289-90.

Giles, Judy. 'Class, Gender and Domestic Consumption in Britain 1920-1950.' In Eds. Emma Casey & Lydia Martens. *Gender and Consumption: Domestic*

Cultures and the Commercialisation of Everyday Life. London: Routledge, 2007.

Gitelman, Lisa. *'Raw Data' Is an Oxymoron*. Cambridge, MA: MIT Press, 2013.

Glander, Timothy. *Origins of Mass Communications Research during the American Cold War: Educational Effects and Contemporary Implications*. London: Lawrence Erlbaum Associates, 1999.

Gleadle, Kathryn. *British Women in the Nineteenth Century*, Basingstoke: Palgrave, 2017.

Gomery, Douglas. *A History of Broadcasting in the United States*. Oxford: Blackwell, 2008.

Gould, Jack. 'Family Life, 1948 A.T. (After Television).' *The New York Times*. (1 August 1948): 12–13.

Grame, Theodore C. *Ethnic Broadcasting in the United States*. American Folklore Center: Library of Congress, 1980.

Green, Venus. *Race on the Line: Gender, Labor, and Technology in the Bell System, 1880–1980*. Durham, NC: Duke University Press, 2001.

Gray, Herman. *Watching Race: Television and the Struggle for 'Blackness.'* Minneapolis: University of Minnesota Press, 1995.

Halper, Donna. *Invisible Stars: A Social History of Women in American Broadcasting*. London: Routledge, 2001.

Halper, Donna. 'Speaking for Themselves: How Radio Brought Women into the Public Sphere.' In Ed. Michael C. Keith. *Radio Cultures: The Sound Medium in American Life*. New York: Peter Lang, 2008.

Hand Ludlow, John. *Television Program Preferences of Listeners in Utah County, Utah*. PhD. Ohio: The Ohio State University, 1957.

Haney, David Paul. *The Americanization of Social Science: Intellectuals and Public Responsibility in the Post-War United States*. Philadelphia, PA: Temple University Press, 2008.

Hansen, Miriam. *Babel & Babylon: Spectatorship in American Silent Film*. Cambridge, MA: Harvard University Press, 2009.

Haralovich, Mary Beth. 'Sitcoms and Suburbs: Positioning the 1950s Homemaker.' *Quarterly Review of Film & Television*. 11, no. 1 (1989): 61–83.

Haralovich, Mary Beth & Rabinovitz, Lauren. *Television, History, and American Culture: Feminist Critical Essays*. Durham, NC: Duke University Press, 1999.

Harper, Richard. *Inside the Smart Home*. New York: Springer, 2006.

Hatch, Kristen. 'Selling Soap: Post-War Television Soap Opera and the American Housewife.' In Ed. Janet Thumin. *Small Screens, Big Ideas: Television in the 1950s*. London: I.B. Tauris, 2002.

Hathaway, Kenneth A. *Television: A Practical Treatise on the Principles upon Which the Development of Television Is Based*. Chicago, IL: American Technical Society, 1933.

Herbert, Stephen. *Eadweard Muybridge: The Kingston Museum Bequest*. East Sussex: The Projection Box, 2004.

Herbert, Stephen. *A History of Early Television: Volume 1*. London: Routledge, 2004a.

Herbert, Stephen. *A History of Early Television: Volume 2*. London: Routledge, 2004b.

Herbert, Stephen. *A History of Early Television: Volume 3*. London: Routledge, 2004c.

Herzog, Herta. 'What Do We Really Know about Daytime Serial Listeners?' In Eds. Frank N. Stanton & Paul F. Lazarsfeld. *Radio Research, 1942–1943*. New York: Duell, Sloan and Pearce, 1944.

Hettinger, Herman S. *A Decade of Radio Advertising*. Chicago, IL: University of Chicago Press, 1933.

Hill, Erin. *Never Done: A History of Women's Work in Media Production*. New Brunswick, NJ: Rutgers University Press, 2016.

Hilmes, Michele. *Radio Voices: American Broadcasting, 1922–1952*. Minneapolis, MN: University of Minnesota Press, 1997.

Hilmes, Michele. 'Desired and Feared: Women's Voices in Radio History.' In Eds. Mary Beth Haralovich & Lauren Rabinovitz. *Television, History, and American Culture: Feminist Critical Essays*. Durham, NC: Duke University Press, 1999.

Hilmes, Michele. *Only Connect: A Cultural History of Broadcasting in the United States*. Wadsworth: Cengage Learning, 2013.

Hoerschelmann, Olaf. *Rules of the Game: Quiz Shows and American Culture*. New York: State University of New York Press, 2006.

Hollingworth, Harry. *Advertising and Selling: Principles of Appeal and Response*. New York & London: D. Appleton and Co, 1913.

Holmes, Su. 'Revisiting Play School: A Historical Case Study of the BBC's Address to the Pre-School Audience.' *The Journal of Popular Television*. 4, no. 1 (2016): 29–47.

Hosley, David H. & Yamada, Gayle K. *Hard News: Women in Broadcast Journalism*. Westport: Connecticut: Greenwood Press, 1987.

Hubbell, Robinson W. *4,000 Years of Television*. London: G. Harrap & Sons, 1946.

Hunt, Verity. 'Electric Leisure: Late Nineteenth-Century Dreams of a Remote Viewing by "Telectroscope".' *Journal of Literature and Science.* 7, no. 1 (2014): 55–76.

Hurwitz, Donald. *Audience Research in American Broadcasting: The Early Years.* Association for Education in Journalism and Mass Communication Annual Convention. August, Oregon, US, 1983.

Huyssen, Andreas. *After the Great Divide: Modernism, Mass Culture, Postmodernism.* Bloomington, IN: Indiana University Press, 1986.

Igo, Sarah. *The Averaged American: Surveys, Citizens, and the Making of a Mass Public.* Cambridge, MA: Harvard University Press, 2008.

Inglis, Ruth. *The Window in the Corner: A Half-Century of Children's Television.* London: Peter Owen, 2003.

Irvin, Richard. *The Early Shows: A Reference Guide to Network and Syndicated Prime-Time Television Series from 1944 to 1949.* Albany, Georgia: BearManor Media, 2018.

Irwin, Mary. 'What Women Want on Television: Doreen Stephens and BBC Television Programmes for Women, 1953–1964.' *Westminster Papers in Communication and Culture.* 8, no. 3 (2011): 99–122.

Jenkins, Charles F. *Vision by Radio: Radio Photographs.* Washington, DC: National Capital Press, 1925.

Jenkins, Charles F. *The Boyhood of an Inventor.* Washington, DC: National Capital Press, 1931.

Johnston, Patricia A. *Real Fantasies: Edward Steichen's Advertising Photography.* Berkeley and Los Angeles, California: University of California Press, 2000.

Jones, Allan. 'Mary Adams and the Producer's Role in Early BBC Science Broadcasts.' *Public Understanding of Science.* 21, no. 8 (2011): 968–83.

Jordanova, Ludmilla. *Sexual Visions: Images of Science and Medicine between the Eighteenth and Twentieth Centuries.* Madison, WI: University of Wisconsin Press, 1989.

Joyrich, Lynne. 'All That Heaven Allows: TV Melodrama, Postmodernism, and Consumer Culture.' In Eds. Lynn Spigel & Denise Mann. *Private Screening: Television and the Female Consumer.* Minneapolis: University of Minnesota Press, 1992.

Kaufman, Helen J. 'The Appeal of Specific Daytime Serials.' In Eds. Frank N. Stanton & Paul F. Lazarsfeld. *Radio Research, 1942–1943.* New York: Duell, Sloan and Pearce, 1944.

Keeler, Amanda. '"A Certain Stigma" of Educational Radio: Judith Waller and "Public Service" Broadcasting.' *Critical Studies in Media Communication.* 34, no. 5 (2017): 495–508.

Kellogg, Halsey D. & Walters, Abner G. 'How to Reach Housewives Most Effectively.' *Broadcasting*. (15 April 1932): 7 & 30.

Kerby, Philip. *The Victory of Television*. New York and London: Harper & Brothers Publishers, 1939.

Kibler, Alison. *Rank Ladies: Gender and Cultural Hierarchy in American Vaudeville*. Chapel Hill: University of North Carolina Press, 2005.

Kingson, Walter K. 'Measuring the Broadcast Audience.' *The Quarterly of Film, Radio and Television*. 7, no. 3 (1953): 291–303.

Kitch, Caroline. *The Girl on the Magazine Cover: The Origins of Visual Stereotypes in American Mass Media*. Chapel Hill, NC: University of North Carolina Press, 2009.

Lavin, Marilyn. 'Creating Consumers in the 1930s: Irna Phillips and the Radio Soap Opera.' *Journal of Consumer Research*. 22, no. 1 (1995): 75–89.

Lazarsfeld, Paul F. *Radio and the Printed Page*. New York: Duell, Sloan and Pearce, 1940.

Lazarsfeld, Paul F. & Kendell, Patricia L. *Radio Listening in America: The People Look at Radio- Again*. New York: Prentice-Hall, 1948.

Lears, Jackson. *Fables of Abundance: A Cultural History of Advertising in America*. New York: Basic Books, 1994.

Levine, Elana. *Her Stories: Daytime Soap Opera and US Television History*. Durham, NC: Duke University Press, 2020.

Lind, Rebecca Ann. 'Understanding the Historical Context of Race and Gender in Electronic Media.' In Ed. Donald G. Godfrey. *Methods of Historical Analysis in Electronic Media*. London: Lawrence Erlbaum Publishing, 2006.

Logan, Anne. 'Gender, Radio Broadcasting and the Role of the Public Intellectual: The BBC Career of Margery Fry, 1928–1958.' *Historical Journal of Film, Radio and Television*. 40, 2 (2020): 389–406.

Lohan, Maria & Faulkner, Wendy. 'Masculinities and Technologies: Some Introductory Remarks.' *Men and Masculinities*. 6, no. 4 (2004): 319–29.

Lumley, Frederick. *Measurement in Radio*. Columbus, OH: Ohio State University, 1934.

Lysack, Krista. *Come Buy, Come Buy: Shopping and the Culture of Consumption in Victoria Women's Writing*. Athens, OH: Ohio University Press, 2008.

Magoun, Alexander. *Television: The Life Story of a Technology*. Westport, CT: Greenwood Press, 2007.

Marcellus, Jane. *Business Girls & Two-Job Wives: Emerging Media Stereotypes of Employed Women*. Cresskill, NJ: Hampton Press, 2011.

Marchand, Roland. *Advertising the American Dream: Making Way for Modernity, 1920–1940*. Berkeley, CA: University of California Press, 1985.

Marks, Patricia. *Bicycles, Bangs, and Bloomers: The New Woman in the Popular Press*. Lexington, KY: The University Press of Kentucky, 2014.

Mayne, Judith. *Cinema and Spectatorship*. London: Routledge, 2002.

McCarthy, Anna. *Ambient Television: Visual Culture and Public Space*. Durham, NC: Duke University Press, 2001.

McGovern, Charles F. *Sold American: Consumption and Citizenship, 1890–1945*. Chapel Hill: The University of North Carolina Press, 2009.

McLean, Donald F. *Restoring Baird's Image*. London: The Institute of Electrical Engineers, 2000.

Meehan, Eileen. 'Ratings and the Institutional Approach: A Third Answer to the Commodity Questions.' *Critical Studies in Mass Communication*. 1, no. 2 (1984): 216–25.

Meehan, Eileen. 'Why We Don't Count: The Commodity Audience.' In Ed. Patricia Mellencamp. *Logics of Television: Essays in Cultural Criticism*. Bloomington and Indianapolis: Indiana University Press, 1990.

Meehan, Eileen. 'Heads of Households and Ladies of the House: Gender, Genre, and Broadcast Ratings, 1929–1990.' In Eds. William Samuel Solomon & Robert Waterman McChesney. *Ruthless Criticism: New Perspectives in U.S. Communication History*. Minneapolis and London: University of Minnesota Press, 1993.

Meehan, Eileen. 'Gendering the Commodity Audience: Critical Media Research, Feminism, and Political Economy.' In Eds. Meenakshi Gigi Durham & Douglas M. Kellner. *Media and Cultural Studies: Keyworks*. Oxford: Blackwell, 2009.

Mellencamp, Patricia. *High Anxiety: Catastrophe, Scandal, Age & Comedy*. Bloomington and Indianapolis: Indiana University Press, 1992.

Mercer, David. *The Telephone: The Life Story of a Technology*. Westport, CT: Greenwood Press, 2006.

Meyers, Cynthia B. 'The Problem with Sponsorship in US Broadcasting, 1930s–1950s: Perspectives from the Advertising Industry.' *Historical Journal of Film, Radio and Television*. 31, no. 3 (2011): 355–72.

Milkman, Ruth. 'Redefining "Women's Work": The Sexual Division of Labor in the Auto Industry during World War II.' In Ed. Nancy F. Cott. *Industrial Wage Work*. London: K.G. Saur, 2013.

Mizejewski, Linda. *Ziegfeld Girl: Image and Icon in Culture and Cinema*. London: Duke University Press, 1999.

Moores, Shaun. "'The Box on the Dresser": Memories of Early Radio and Everyday Life.' *Media, Culture & Society*. 10, no. 1 (1988): 23–40.

Moores, Shaun. *Media/Theory: Thinking about Media and Communications.* London and New York: Routledge, 2007.

Mullen, Megan. *The Rise of Cable Programming in the United States: Revolution or Evolution?* Austin, TX: University of Texas Press, 2003.

Mulvey, Laura. 'Visual Pleasure and Narrative Cinema.' *Screen*. 16, no. 3 (1975): 6–18.

Murphy, Catherine. 'On an Equal Footing with Men?' *Women and Work at the BBC, 1922–1939*. PhD Thesis. Goldsmiths College: University of London, 2011.

Murphy, Kate. *Behind the Wireless: A History of Early Women at the BBC.* Basingstoke: Palgrave, 2016.

Murphy, Kate. "'New and Important Careers': How Women Excelled at the BBC, 1923–1939.' *Media International Australia*. 161, no. 1 (2016): 18–27.

Murphy, Kate. 'Relay Women: Isa Benzie, Janet Quigley and the BBC's Foreign Department, 1930–38.' *Feminist Media Histories*. 5, no. 3 (2019): 114–39.

Murray, Matthew. 'Matinee Theater: Difference, Compromise and the 1950s Daytime Audience.' In Ed. Janet Thumim. *Small Screen, Big Ideas: Television in the 1950s*. London: I.B. Tauris, 2002, 131–48.

Murray, Susan. *Hitch Your Antenna to the Stars: Early Television and Broadcast Stardom.* London: Routledge, 2005.

Musser, Charles. *The Emergence of Cinema: The American Screen to 1907*, Volume 1. Berkeley, CA: University of California Press, 1994.

Musser, Charles. 'At the Beginning: Motion Picture Production, Representation and Ideology at the Edison and Lumiére Companies.' In Eds. Lee Grieveson & Peter Krämer. *The Silent Cinema Reader*. London: Routledge, 2004.

Napoli, Philip M. *Audience Evolution: New Technologies and the Transformation of Media Audiences*. New York: Columbia university Press, 2011.

Nava, Mica. 'Modernity's Disavowal: Women, the City and the Department Store.' In Eds. Mica Nava & Alan O'Shea. *Modern Times: Reflections on a Century of English Modernity*. London: Routledge, 1996.

Newcomb, Horace. *Encyclopedia of Television*. Second Edition. London and New York: Routledge, 2014.

Norman, Bruce. *Here's Looking at You: The Story of British Television, 1908–1939*. London: Royal Television Society, 1984.

Novotny, Patrick. *The Press in American Politics, 1787–2012*. Santa Barbara, CA: Praeger, 2014.

O'Dell, Cary. *Women Pioneers in Television: Biographies of Fifteen Industry Leaders*. Jefferson, NC: McFarland & Company, 1997.

O'Dell, Cary. 'A Station of Their Own: The Story of the Women's Auxiliary Television Technical Staff (WATTS) in World War II Chicago.' *Television Quarterly*. 30, no. 2 (2000): 58–67.

O'Dell, Cary. *June Cleaver Was a Feminist!: Reconsidering the Female Characters of Early Television*. Jefferson, MC: McFarland & Company, 2013.

O'Dell, Cary. 'Thelma Prescott, Television's First Female Director/Producer.' 2017. Available at: https://blogs.loc.gov/now-see-hear/2017/05/thelma-prescott-televisions-first-female-producerdirector/ [Accessed on 12 December 2017].

Oakley, Ann. *Woman's Work: The Housewife Past and Present*. New York: Vintage Books, 1974.

Oldenziel, Ruth. *Making Technology Masculine: Men, Women and Modern Machines in America, 1870–1945*. Amsterdam: Amsterdam University Press, 1999.

Ortner, Sherry B. 'Is Female to Male as Nature Is to Culture?' *Feminist Studies*. 1, no. 2 (1972): 5–31.

Ozmun, David. 'Opportunity Deferred: A 1953 Case Study of a Woman Working in Network Television News.' *Journal of Broadcasting & Electronic Media*. 52 (2008): 1–15.

Pandora, Katherine. 'The Permissive Precincts of Barnum's and Goodrich's Museums of Miscellaneity: Lessons in Knowing Nature for New Learners.' In Eds. Carin Berkowitz & Bernard Lightman. *Science Museums in Transition: Cultures of Display in Nineteenth-Century Britain and America*. Pittsburgh, PA: University of Pittsburgh Press, 2017.

Parlin, Charles Coolidge. *The Merchandising of Textiles*. Philadelphia, PA: Curtis Publishing Co, 1914.

Parrett, Catriona. *More Than Mere Amusement: Working-Class Women's Leisure in England, 1750–1914*. Boston, MA: Northeastern University Press, 2001.

Paxman, Andrew. 'The Rise of U.S. Spanish-Language Radio.' *Journalism History*. 44, no. 3 (2018): 174–86.

Peiss, Kathy. 'American Women and the Making of Modern Consumer Culture.' *The Journal for Multi Media History*. 1, no. 1 (1998).

Available at: https://www.albany.edu/jmmh/vol1no1/peiss-text.html [Accessed on 12 December 2017].

Peiss, Kathy. *Cheap Amusements: Women and Leisure in Turn-of-the-Century New York.* Philadelphia, PA: Temple University Press, 2011.

Petro, Patrice. 'Mass Culture and the Feminine: The "Place" of Television in Film Studies.' *Cinema Journal.* 25, no. 3 (1986): 5–21.

Radway, Janice. *Reading the Romance: Women, Patriarchy, and Popular Culture.* Chapel Hill and London: University of North Carolina Press, 1991.

Rappoport, Leon. *Punchlines: The Case for Racial, Ethnic, and Gender Humor.* London: Praeger, 2005.

Reekie, Gail. *Temptations: Sex, Selling and the Department Store.* Sydney: Allen & Unwin, 1993.

Reilly, William J. *Marketing Investigations.* New York: The Ronald Press Company, 1929.

Robida, Albert. 1883. *The Twentieth Century.* Trans. Philippe Willems. Ed. Arthur B. Evans. Middletown, CT: Wesleyan University Press, 2004.

Rosa-Salas, Marcel. 'Making the Mass White: How Racial Segregation Shaped Consumer Segmentation.' In Eds. Guillaume D. Johnson, Kevin D. Thomas & Anthony Kwame Harrison. *Race in the Marketplace: Crossing Critical Boundaries.* Basingstoke: Palgrave, 2019.

Rutherford, Williams. Janice. *Selling Mrs. Consumer: Christine Frederick & the Riske of Household Efficiency.* Athens and London: University of Georgia Press, 2010.

Ryan, Mary P. *Cradle of the Middle-Class: The Family in Oneida County, New York, 1790–1865.* Cambridge: Cambridge University Press, 1981.

Sandon, Emma. 'Engineering Difference: Women's Accounts of Working as Technical Assistants in the BBC Television Service between 1946 and 1955.' *Feminist Media Histories.* 4, no. 4 (2018): 8–32.

Sassatelli, Roberta. *Consumer Culture: History, Theory and Politics.* London: SAGE, 2007.

Scanlon, Jennifer. *Inarticulate Longings: The Ladies' Home Journal, Gender, and the Promises of Consumer Culture.* London and New York: Routledge, 1995.

Scanlon, Jennifer. 'Introduction.' In Ed. Jennifer Scanlon. *The Gender and Consumer Culture Reader.* New York and London: New York University Press, 2000.

Schwarzkopf, Stefan. 'The Statisticalization of the Consumer in British Market Research, *c.* 1920–1960: Profiling a Good Society.' In Eds. Tom Crock &

Glen O'Hara. *Statistics and the Public Sphere: Numbers and the People in Modern Britain, c. 1800–2000.* London: Routledge, 2011.

Schwarzkopf, Stefan. 'The Politics of Enjoyment: Competing Audience Measurement Systems in Britain, 1950–1980.' In Eds. Jerome Bourdon & Cecile Meadal. *Television Audiences around the World: Deconstructing the Ratings Machine.* New York: Palgrave Macmillan, 2014.

Schwarzkopf, Stefan. 'In Search of the Consumer: The History of Market Research from 1890 to 1960.' In Eds. D.G. Brian Jones & Mark Tadajewski. *The Routledge Companion to Marketing History.* London: Routledge, 2016.

Schweitzer, Marlis. *When Broadway Was the Runway: Theatre, Fashion, and American Culture.* Philadelphia, PA: University of Pennsylvania Press, 2009.

Scott, Anne Firor. *Natural Allies: Women's Associations in American History.* Urbana and Chicago: University of Illinois Press, 1992.

Seger, Linda. *When Women Called the Shots: The Developing Power and Influence of Women in Television and Film.* New York: Henry Holt, 1996.

Sewell, Jessica Ellen. *Women and the Everyday City: Public Space in San Francisco, 1890–1915.* Minneapolis, MN: University of Minnesota Press, 2011.

Sewell, Philip W. *Television in the Age of Radio: Modernity, Imagination, and the Making of a Medium.* New Brunswick, NJ: Rutgers University Press, 2014.

Shiers, George & Shiers, May. *Early Television: A Bibliographic Guide to 1940.* New York and London: Garland Publishing, 1997.

Silvey, Robert. *Who's Listening? The Story of BBC Audience Research.* London: Allen & Unwin, 1974.

Sioussat, Helen. *Mikes Don't Bite.* New York: L.B. Fischer, 1943.

Slotten, Hugh R. *Radio and Television Regulation: Broadcast Technology in the United States, 1920–1960.* London: Johns Hopkins University Press, 2000.

Smythe, Dallas. 'Communications: Blindspot of Western Marxism.' *Canadian Journal of Political and Social Theory.* 1, no. 3 (1977): 1–27.

Spaulding, Stacy. 'Did Women Listen to News? A Critical Examination of Landmark Radio Audience Research (1935–1948).' *Journalism & Mass Communication Quarterly.* 82, no. 1 (2005): 44–61.

Spaulding, Stacy. 'Lisa Sergio's "Column of the Air": An Examination of the Gendered History of Radio (1940–1945).' *American Journalism.* 22, no. 1 (2005): 35–60.

Spigel, Lynn. 'Installing the Television Set: Popular Discourses on Television and Domestic Space, 1948–1955.' *Camera Obscura.* 6, no. 1 (1988): 9–46.

Spigel, Lynn. 'The Domestic Economy of Television Viewing in Postwar America.' *Critical Studies in Mass Communication.* 6, no. 4 (1989): 337–54.

Spigel, Lynn. *Make Room for TV: Television and the Family Ideal in Postwar America.* Chicago, IL: The University of Chicago Press, 1992.

Stamp, Shelley. *Movie-Struck Girls: Women and Motion Picture Culture after the Nickelodeon.* Princeton, NJ: Princeton University Press, 1998.

Stanley, Autumn. *Mothers and Daughters of Invention: Notes for a Revised History of Technology.* New Brunswick, NJ: Rutgers University Press, 1995.

Stanton, Andrea L. 'Television.' In Eds. Andrea L. Stanton, Edward Ramsamy, Peter J. Seybolt & Carolyn M. Elliott. *Cultural Sociology of the Middle East, Asia, and Africa: An Anthology.* London: SAGE, 2012.

Stanton, Frank N. & Lazarsfeld, Paul F. *Radio Research, 1942–1943.* New York: Duell, Sloan and Pearce, 1944.

Stanton, Michael Joseph. *A History of the Research and Planning Department of the National Broadcasting Company, Incorporated (1931 to 1976).* PhD Thesis. Ann Arbor, MI: University Microfilms International, 1977.

Starch, Daniel. *Advertising: Its Principles, Practice, and Technique.* Chicago, IL: Foresman and Co., 1914.

Sterling, Christopher H. & Kittross, John Michael. *Stay Tuned: A History of American Broadcasting.* London: Lawrence Erlbaum Associates, 2009.

Stewart, David W. 'The Evolution of Market Research.' In Eds. Pauline Maclaran, Michael Saren, Barbara Stern & Mark Tadajewski. *The SAGE Handbook of Marketing Theory.* London: SAGE, 2010.

Stivers, Richard. *Technology as Magic: The Triumph of the Irrational.* New York: Continuum, 2001.

St John, Jacqueline D. 'Sex Role Stereotyping in Early Broadcast History: The Career of Mary Margaret McBride.' *Frontiers: A Journal of Women Studies.* 3, no. 3 (1978): 31–8.

Stole, Inger L. 'Televised Consumption: Women, Advertisers and the Early Daytime Television Industry.' *Consumption, Markets and Culture.* 6, no. 1 (2003): 65–80.

Street, Sean. 'BBC Sunday Policy and Audience Response: 1930–1945.' *Journal of Radio Studies.* 7, no. 1 (2000): 161–79.

Summerfield, Penny. 'It Did Me Good in Lots of Ways': British Women in Transition from War to Peace.' In Eds. Claire Duchen & Irene Bandhauer-Schoffmann. *When the War Was Over: Women, War, and Peace in Europe, 1940–1956.* London: Leicester University Press, 2000.

Swiencicki, Mark A. 'Consuming Brotherhood: Men's Culture, Style and Recreation as Consumer Culture, 1880–1930.' In Ed. Lawrence B. Glickman. *Consumer Society in American History: A Reader*. Ithaca, NY: Cornell University Press, 1999.

Terkanian, Kathryn. *Women, Work, and the BBC: How Wartime Restrictions and Recruitment Woes Reshaped the Corporation, 1939–45*. PhD Thesis. Bournemouth University, 2019.

Thumim, Janet. *Inventing Television Culture: Men, Women, and the Box*. Oxford: Oxford University Press, 2004.

Turnock, Rob. *Television and Consumer Culture: Britain and the Transformation of Modernity*. London: I.B. Tauris, 2007.

Vaillant, Derek W. 'Sounds of Whiteness: Local Radio, Racial Formation, and Public Culture in Chicago, 1921–1935.' *American Quarterly*. 54, no. 1 (2002): 25–6.

Veblen, Thorstein. 1899. *The Theory of the Leisure Class*. Dover Thrift Edition. Mineola, NY: Dover Publications, 2016.

von Schilling, James A. *The Magic Window: American Television 1939–1953*. New York: Haworth Press, 2013.

Wajcman, Judy. *Feminism Confronts Technology*. Cambridge: Polity Press, 1991.

Wajcman, Judy. 'Addressing Technological Change: A Challenge to Social Theory.' In Eds. Morton Winston & Ralph Edelbach. *Society, Ethics, and Technology*. Fifth Edition. Boston, MA: Cengage Learning, 2013.

Ward, Douglas. *A New Brand of Business: Charles Coolidge Parlin, Curtis Publishing Company, and the Origins of Market Research*. Philadelphia, PA: Temple University Press, 2009.

Ware, Susan. *Still Missing: Amelia Earhart and the Search for Modern Feminism*. New York: W. W. Norton, 1993.

Ware, Susan. *It's One O'Clock and Here Is Mary Margaret McBride: A Radio Biography*. New York: New York University Press, 2005.

Weil Davis, Simone. *Living up to the Ads: Gender Fictions in the 1920s*. Durham, NC: Duke University Press, 2000.

Weinstein, David. *The Forgotten Network: DuMont and the Birth of American Broadcasting*. Philadelphia, PA: Temple University Press, 2006.

Welter, Barbara. *Dimity Convictions: The American Woman in the Nineteenth Century*, Athens, OH: Ohio University Press, 1976.

Wheatley, Helen. 'Television in the Ideal Home.' In Eds. Rachel Moseley, Helen Wheatley & Helen Wood. *Television for Women: New Directions.* London: Routledge, 2016.

Wheeler, Leigh Ann. *Against Obscenity: Reform and the Politics of Womanhood in America, 1873–1935.* Baltimore, MD: Johns Hopkins University Press, 2004.

Wodak, Ruth. *Gender and Discourse.* London: SAGE, 1997.

Wood, Helen. 'Television – the Housewife's Choice? The 1949 Mass Observation Television Directive, Reluctance and Revision.' *Media History.* 21, no. 3 (2015): 342–59.

Zakharine, Dmitri. 'Voice- E-Voice-Design- E-Voice-Community: Early Public Debates about the Emotional Quality of Radio and TV Announcers' Voices in Germany, the Soviet Union and the USA (1920–1940).' In Eds. Dmitri Zakharine & Nils Meise. *Electrified Voices: Medial, Socio-Historical and Cultural Aspects of Voice Transfer.* Gottingen: V & R, 2013.

Index

www.ingramcontent.com/pod-product-compliance
Lightning Source LLC
Chambersburg PA
CBHW060151280326
41932CB00012B/1723